Routledge Revivals

Changing Resource Problems of the Fourth World

Climbing food, fertiliser and mineral prices as well as the Arab oil embargo in the seventies had severe economic consequences in developing countries. Originally published in 1976, this study explores the effects of these developments in the fourth world and how they can adjust to an international economy with a particular focus on resource availability in terms of energy and agriculture. This title will be of interest to students of Environmental Studies.

Changing Resource Problems of the Fourth World

Edited by
Ronald G. Ridker

First published in 1976
by Resources for the Future, Inc.

This edition first published in 2016 by Routledge
2 Park Square, Milton Park, Abingdon, Oxon, OX14 4RN
and by Routledge
711 Third Avenue, New York, NY 10017

Routledge is an imprint of the Taylor & Francis Group, an informa business

Publisher's Note
The publisher has gone to great lengths to ensure the quality of this reprint but points out that some imperfections in the original copies may be apparent.

Disclaimer
The publisher has made every effort to trace copyright holders and welcomes correspondence from those they have been unable to contact.

A Library of Congress record exists under LC control number: 75042978

ISBN 13: 978-1-138-95537-0 (hbk)
ISBN 13: 978-1-315-66638-9 (ebk)

CHANGING RESOURCE PROBLEMS OF THE FOURTH WORLD

Ronald G. Ridker, *Editor*

Resources for the Future
Washington, D.C.

February 1976

RFF WORKING PAPER PD-1
DISTRIBUTED BY THE JOHNS HOPKINS UNIVERSITY PRESS

RESOURCES FOR THE FUTURE, INC.
1755 Massachusetts Avenue, N.W., Washington, D.C. 20036

Resources for the Future is a nonprofit organization for research and education in the development, conservation, and use of natural resources and the improvement of the quality of the environment. It was established in 1952 with the cooperation of the Ford Foundation. Part of the work of Resources for the Future is carried out by its resident staff; part is supported by grants to universities and other non-profit organizations. Unless otherwise stated, interpretations and conclusions in RFF publications are those of the authors; the organization takes responsibility for the selection of significant subjects for study, the competence of the researchers, and their freedom of inquiry. This book is a product of RFF's Institutions and Public Decisions Division, directed by Clifford S. Russell.

This material has been published as received without the usual editing and type-setting in order to speed its distribution.

Library of Congress Catalog Card Number 75-42978
ISBN 0-8018-1847-8

Manufactured in the United States of America

RFF Working Paper PD-1

TABLE OF CONTENTS

I. INTRODUCTION

Ronald G. Ridker

Since 1972 the developing countries have experienced a series of economic shocks as severe as any encountered during World War II. After a period of relative stability, world food and fertilizer prices began their dramatic climb. In 1973, minerals and metals prices followed suit. Then came the Arab oil embargo, a four-fold increase in oil prices and a severe cutback in fertilizer exports by major world producers. A few developing countries, those with oil, minerals, metals and timber to sell, experienced windfall gains (some only briefly); but the majority of poor countries -- the vast majority if judged by the size of their populations -- have seen their terms of trade and as a consequence their balance of payments position deteriorate with the impact of a bursting dam. While some of these shocks are likely to prove temporary -- some prices have already backed away from their 1974 highs -- the most serious problems for these countries are posed by the situation in the petroleum and food markets where relief is unlikely to come soon.

What are the consequences of these developments for the Fourth World, the poorest of the poor? How can they adjust to these new facts of international economic life? Can they put together OPEC-type organizations to reverse or at least stem the tide of market forces in any of their export markets? To what extent can they find substitutes on either of the supply or demand side for commodities such as oil and fertilizer?

In particular, what adjustments in the critical agricultural sector make sense in order to minimize the welfare losses involved? How can research help in the search for alternatives?

While no one has adequate answers to these questions, the four papers included in this volume provide insights and background materials that are important for serious students of these problems. By presenting them in RFF's working paper series, we hope they will receive broader circulation and consideration than they might in isolation from one another.

The first paper, by Stern and Tims on "The Relative Bargaining Strength of the Developing Countries", includes one of the most comprehensive treatments available on the question of whether more OPEC-like cartels are likely. For each export commodity of importance to developing countries the authors consider the principal factors determining price and income elasticities of demand supply and how these are likely to change over time. Their conclusions leave little hope that answers for developing countries' problems can be found through collective actions: first, there is no commodity that approaches oil in terms of its universal essentiality, lack of readily available substitutes and the capacity of a few producers to restrict output without impairing their own import, employment or output position; and second, unrestricted market forces are more likely to reduce than to raise the relative prices of many primary commodities in the next few years. In the light of recent developments in international forums, the authors may have given too little weight to political elements in the equation; but if they are correct about the underlying economic forces, the fruits of collective actions on the part of many primary producer groups are likely to turn sour rather quickly.

The second paper on "Domestic Adjustments and Accommodations to Higher Raw Material and Energy Prices" by T. L. Sankar finds the prospects for maneuver domestically to be as limited as they are internationally, during the next decade or so. Reliance on international reserves, borrowing and foreign assistance are at best temporary expedients and ones that are not open to all poor countries. Conservation measures cannot be expected to accomplish much given already low per capita consumption figures. Substitution of indigenous for imported fuels and minerals is a possibility for some countries but in most cases the substitute processes are more capital intensive, require more sophisticated machinery and involve long gestation lags. For example, India can substitute coal for oil in rail transport, but this appears to be economically justifiable only on lines where it makes sense to electrify; a transformation that requires considerable capital and time.

Finally, an effort can be made to alter the economic structure in such a way that consumption of the more expensive imported materials does not increase with aggregate output in the way it has done in the past. Alan Strout, in his paper on "Energy and the Less Developed Countries: Needs for Additional Research", considers what these typical patterns have been for energy as development proceeds and uses them to show the frightening dimensions of the problem facing many developing countries, assuming future petroleum prices remain at their current levels. Sankar suggests a few possibilities for altering these patterns -- for example, slowing down mechanization of agriculture and increasing the value added of exports through more processing of primary products particularly where such additional processing is labor intensive -- but he does not hold out much hope that such deviations from historic patterns of development can be very substantial. Sizeable reductions in consumption and investment over what they might otherwise have been seems inevitable.

By looking more specifically at "Energy Use and Agricultural Production in Developing Areas" Kenneth Frederick makes these results even more concrete. He finds that while developing countries have great potential for increasing agricultural production, that potential can be achieved during the next decade or two only by significantly increasing the use of fertilizer, pesticides and energy, making these countries more dependent on imported inputs. In the longer run, biological engineering may reduce dependence on chemical fertilizers and pesticides; new methods of cultivation and irrigation may reduce fuel requirements for tillage and pumping; and agricultural waste products can be used to produce biogas on site. But many of these methods are capital intensive (both directly and in terms of supporting social overheads), some require additional years of fundamental research and all depend on inducing millions of isolated and poorly educated peasants to change their traditional patterns of behavior. In the meantime, the dependence on petro-chemicals will continue to grow.

Strout's paper generalizes this conclusion by studying the implications for individual developing countries of assuming that their growth paths follow the long-term pattern described by cross-country comparisons. Given this assumption, a large developing country would have to increase its per capita commercial energy consumption by a factor of ten in order to increase its per capita gross domestic product from $125 to $500. At current energy costs, such a country would have to increase its export earnings by some 77% more than they would normally be at the $500 income level. He then considers the principal ways that energy is used in such economies and concludes that this pattern cannot be significantly altered, at least during the next several decades, without impairing the industrialization process.

All such conclusions are highly tentative, as he and the other authors realize, if for no other reason than that historic growth patterns were established during an era of relatively low energy prices. It is therefore especially fitting that Strout concludes with research suggestions aimed at trying to find alternative development paths for the countries of this Fourth World. The search is likely to be a long one involving forays into fields as disparate as nuclear physics and urban settlement patterns. We can only hope that it begins bearing fruit before the political and social pressures of a deteriorating situation rob these countries of the chance for orderly progress in some new directions.

II. THE RELATIVE BARGAINING STRENGTHS OF THE DEVELOPING COUNTRIES

Ernest Stern and Wouter Tims[1]

Introduction

> "A particular aspect of economic stability to which special attention must be given is the stabilization of world raw material prices.... The unfortunate consequences of large fluctuations in these prices have been particularly evident in the past few years, and it is apparent that the security of a greater measure of stabilization would make a real contribution to balance of payments equilibrium.... While the need for avoiding excessive fluctuations will be readily recognized, there is no easy solution that is equally applicable to all commodities. Each commodity must be approached as a special case so as to take account of its characteristics and of the interests of the countries concerned as sellers and buyers."

This description of the commodity problem could have been written in 1975 but in fact it was written more than two decades ago.[2] The lack of progress since then in dealing with commodity prices -- both to reduce the amplitude of fluctuations and to maintain or increase their prices relative to those of manufactured goods -- has been a frequent source of friction between the industrialized and the developing countries. Efforts to manage commodity prices through international cooperation have not been lacking but, in general, these have not been very successful. The world economic situation and the actions of groups of commodity producing countries have heightened the awareness of raw material producers and consumers alike to the possibilities of supply management. Speculation about the chances for success of such actions obviously must be larded liberally with humility derived from the knowledge that a mere four years ago it was held with certitude that a collective effort to raise oil prices was impossible. Nonetheless, it is worthwhile to set out the interrelationships bearing on the access to, and the prices of, primary commodities so that there will be a framework for discussion of alternative scenarios.

The unilateral increase of oil prices, the general commodity price boom of 1972-74, and the concern with the possible depletion of the world's resources, is sometimes seen as the beginning of a major shift in power relationships between the industrialized and developing countries. Fred Bergaten called attention to the interaction of these factors as early as the summer of 1973[3/] and recently has written:

> "As a result of their shabby treatment in the past, and skepticism about meaningful change in the attitudes of the rich, the countries of the Third World are unlikely to recant quickly the policies based on that new power which have so sharply boosted their pocketbooks and prestige."[4/]

The new power of the commodity exporting countries on the economic side is seen to derive essentially from the control over raw material supplies and the consequent prospect of additional cartel-type action. This leverage is thought to be a basis not only for increasing foreign exchange earnings but also for greater effectiveness in international political forums.

The analysis of bargaining strengths is highly complex and many of the factors at play can only be assessed subjectively. Bargaining is a process in which parties with different interests and conflicting objectives attempt to reach an agreement which benefits the parties in rough proportion to their strengths. When applied to relations between industrialized and developing countries, complications arise since the countries of the world are not neatly divided into two such groups. Instead, they display a spectrum of strengths and interests that spans both groups, whether the criterion is income per capita, exports of primary products, or import dependence. Consequently, wherever this spectrum is divided for analysis there are multiple objectives within each group.

This paper will not explore the dynamics of bargaining power but limit itself to the question of how recent economic changes in commodity markets have affected the economic strengths of the developing countries and how the relevant factors are likely to develop in the future.

Current Bargaining Strength

The generally accepted thesis that countries of the developing world have lacked significant bargaining power generally has been formulated in three dimensions: demand, supply, and degree of essentiality. Assertions to the contrary -- and recent experience related to the world market for crude oil -- are cast largely in terms of the last element, essentiality.

In discussing changes in the relative bargaining strengths of the developing countries, we must consider their role as suppliers and consumers of goods and services. The potential to bargain to advantage obviously is not limited to a significant or dominant position as a supplier of goods and services, or factors of production; the same strength can be derived from a similarly important role on the demand side. The basic conclusions of the demand and supply analysis must then be assessed in terms of the relative importance, or essentiality, of goods and services supplied or demanded.

The best known arguments regarding the lack of bargaining strength of developing countries have been in the trade area.[5] Developing countries depend for their foreign exchange earnings on a small number of primary commodities; these constituted in 1960 more than 90 percent of their exports, and still constituted close to 80 percent in 1972. Their dependence on these commodities for exports is not matched by an equally dominant share in the markets for several of these products (see Table 1). Although it is an oversimplification just to compare totals, it nonetheless is of some significance to note that developing countries' exports of primary commodities were only 40 percent of world primary exports in 1955, and this percentage declined gradually to 35 percent in 1970-1972. (See Table 2).

The declining share in world trade of primary commodities has been exacerbated by sharp fluctuations in output and prices which have given rise to major competition from synthetics and other substitutes. Each period of high prices for primary commodities has led to investments for synthetic substitutes and alternative technologies which not only bring prices down but also tend to reduce raw material requirements per unit of output and to capture market shares virtually permanently.

The position of many developing countries has also been eroded by the reduction of the preferences that once applied to major segments of their exports. Many of these preferences have been reduced or abolished, or their coverage has been restricted.[6] Some developing countries did

Table 1. Market Position in Major Commodities of LDC Exporters (1970–72 Average in Percentages)

	Mineral Exporters				Other Developing Countries with Per Capita Income						All Developing Countries	
	Oil Exporters		Other Mineral Exporters		Above $375		$200–375		Below $200			
	Share in Total Exports of Country Group	Share in World Market	Share in Exports	Market Share	Share in Exports	Market Share	Share in Exports	Market Share	Share in Exports	Market Share	Share in Total LDC Exports	Share in World Market
Oil	84	69	—	—	3	2	2	—	4	1	35	73
Coffee	—	—	2	3	8	60	4	9	9	22	5	97
Copper	—	—	62	48	—	—	3	5	1	1	5	54
Sugar	—	—	2	3	3	22	5	12	13	32	4	71
Cotton	—	—	2	2	2	14	7	17	7	13	3	57
Timber	1	3	—	—	2	10	7	10	1	1	2	26
Iron ore	1	5	4	12	1	9	3	8	4	10	2	39
Rubber	1	24	—	—	2	52	2	14	1	6	2	98
Cocoa	1	24	—	—	1	17	6	49	1	8	1	100
Beef	—	—	—	—	3	28	—	—	—	—	1	30
Tea	—	—	—	—	—	—	—	—	7	73	1	83
Tin	1	2	1	4	1	42	3	24	—	—	1	86
Bananas	—	—	—	—	2	66	2	20	—	—	1	92
Other primary commodities	1	1	15	2	9	6	14	3	7	2	8	15
Other exports	10	1	12	—	63	6	42	1	45	2	29	9
Total	100		100		100		100		100		100	
Export weighted market shares		59		31		14		8		14		49
All of the above, excluding oil		—		—		—		—		—		35

Source: International Bank for Reconstruction and Development (1974).
Note: Column 1 under each country category shows the share of a particular commodity in the total export basket of the group; column 2 under each group shows the share of the group's exports in total world trade in the commodity.

This table is reproduced from the American Journal of Agricultural Economics, Vol. 57, No. 2, p. 226, May 1975 with permission of the editor.

TABLE 2

DEVELOPING COUNTRIES' SHARE OF WORLD EXPORTS OF PRIMARY COMMODITIES FOR SELECTED YEARS

Years	LDC Exports (US $ billion)	World Exports (US $ billion)	LDC Share of World Exports (Percent)
1955	21.91	54.12	40.5
1960	24.80	67.89	36.5
1965	32.04	90.03	35.6
1970	45.48	132.32	34.4
1971	49.46	142.18	34.8
1972	57.64	166.07	34.7

ces: UNCTAD, Handbook of International Trade and Development Statistics, 1972; U.N. Monthly Bulletin of Statistics, September 1974.

benefit from new preferential arrangements in some of their markets, but these have been the exceptions rather than the rule. At the same time non-tariff barriers on primary products have grown over the past 10-20 years even though, on average, tariffs may have remained the same or have been reduced. This has exposed an increasing share of the developing countries' export trade to international competition and to restrictions of their access to markets, weakening their bargaining strength.

Prices have become less predictable and more volatile in recent years as a consequence of movement toward more flexible exchange rates. The rapid rates of inflation in the industrialized countries are being passed along more fully into the export sectors than in the past. This has a particularly strong impact on the developing countries because they are price takers in their import markets. While some developing countries may benefit from higher export prices for primary commodities, developing countries' imports account for only a small share of the trade in primary commodities and, generally, cannot affect price. For some strategic commodities, such as foodgrains, their position is eroded further by the wide annual fluctuations in domestic production. Their imports of manufactured goods and services also constitute a small share of world trade though these goods dominate the imports baskets of the developing countries.

The imbalance in relative positions also is discernible on the capital side. In the 1960's two-thirds of the net capital inflows of developing countries were obtained on concessional terms from official sources; now less than one-third is obtained from these sources. Part of this shift reflects the substantial growth achievement of a few countries who now can rely on the capital markets for finance. But other countries in this group, given their per capita incomes, still would be entitled to concessional terms on equity grounds if the supply of concessional capital were not unilaterally determined and hence highly limited. Today such capital is inadquate to meet the needs of even the poorest countries. In the area of private investment, countries must deal with corporations which have a multinational network of operations and which often are more effective negotiators than the host countries which may have fewer alternatives than the corporations. The developing countries also are in a worse position because of their dependence on imported technology which is associated with the capital inflows. The negative position toward foreign investment that developing countries increasingly have adopted is less a sign of strength than a reflection of weakness. Afraid of striking a poor bargain, and uncertain about their capacity to strike a good one, they prefer not to bargain at all or to bargain only in carefully selected areas.

Even prior to the increase in the price of crude oil by OPEC in late 1973, there was considerable discussion of the capacity of primary commodity producing countries to improve the conditions of their markets. The argument which suggested a stronger position of the developing countries than was traditionally held was based essentially on alleged long-term supply constraints. In the case of minerals and metals (including oil) it was pointed out that the resources of the world are exhaustible and that the rapid growth of production and consumption was bringing the day nearer when absolute limits on production capacity, or even production declines, would improve the market strength of the major suppliers.[7] Similarly, with respect to food -- notably grains but also the resources of the sea -- it was argued that there are limits to the production capacity of the world and to the rate at which remaining land suitable for agriculture can be developed.

The Market Outlook

Prices of primary commodities depend on factors which are mostly outside the control of the developing countries exporting these commodities. The growth of demand and of the production of substitutes (both natural and synthetic), the changes in protective measures in importing countries and more recently the instability associated with high rates of inflation and fluctuations of exchange rates all belong to this category of largely exogenous variables. The producing countries can have an impact on prices and market shares through their control of market supplies and by virtue of their competitiveness, although the latter is blunted by the protective systems in major markets. The outlook for primary commodity prices as presented below does not take account of the possible effects of producers' arrangements, except in the case of oil.

Commodity prices surged in 1973 and 1974, after a period of remarkably stable prices since the Korean War. However, compared to prices of manufactured foods, the prices of primary commodities tended to decline throughout the 'fifties and up to about 1962. Between 1962 and the middle of 1972 there were no major changes in the commodity (net barter) terms of trade. More recently, between 1970 and 1973, the rise in commodity prices was far greater than the increase in the prices of exported manufactured goods. The very rapid growth of demand which accompanied the rise of economic activity in virtually all OECD countries put severe pressures on the supplies of a number of commodities (including minerals, metals and non-food agricultural products), further aggravated by the failure of foodgrain crops and of the Peruvian fish catch. The low levels of investment in new fertilizer capacity during recent years also resulted in supply constraints for this essential product, threatening a continuance of world-wide foodgrain shortages.

In the course of 1974, prices of most primary commodities have, with few exceptions, begun to soften. This is expected to continue throughout 1975 while demand conditions are weak. Crude oil will remain the most notable exception; its price may in the near future, at least in current dollars, rise further in 1975 as compared to 1974. Other possible exceptions, but to a much lesser degree, are cereals and sugar.

Price projections beyond 1975 are hazardous, considering the uncertainties with respect to growth in the industrial countries, general price developments in international trade and the behaviour of exchange rates. The price projections for the longer term presented here are based on the assumption that growth in the industrial countries will, after 1975, move towards its historical long-term trend and that inflation rates measured in US dollars will gradually abate to some 7 percent per year towards the end of the decade.

These assumptions are on the optimistic side, but without those, long-term prospects for the developing countries would be even less favorable. The serious impediments to their growth prospects, described in the following paragraphs, point to the importance for the developing countries of sound growth policies in the OECD countries. The probability of deviations from the projected growth path for the developing countries is to the downward side.

Measured in constant dollars - i.e., measured relative to past and projected price changes for manufactured goods in world trade - most primary product prices (excluding oil) are expected by 1976 to return to their level of the 1960's and to remain there into the early 1980's. Commodities for which the constant dollar prices around 1980 are likely to be higher than in the 'sixties include sugar, livestock products, timber, zinc, bauxite and iron ore, as well as steel and phosphate rock. Commodities for which the constant dollar prices will be significantly lower include tea, fruits, jute, sisal, rice and manganese ore.

These price projections may, if anything, be somewhat optimistic: in addition to the optimistic assumptions regarding economic growth in the industrial countries, primary commodity prices may not keep up as well with general inflation as is assumed. The assessment of prospects does take account of possible medium-term supply bottlenecks, both for some primary commodities and for some synthetic substitutes. Although it is not impossible that long-term constraints exist as well, these do not affect projected prices over the time span observed here.

The projected prices have a differentiated effect by groups of developing countries. In part this reflects the composition of their exports, but for another important part these projections have implications

TABLE 3

Indices of Commodity Prices, 1950-1980
(in constant 1967-69 dollars; 1967-69 = 100)

	Agricultural Products		Metals, Minerals	Petroleum	Total (excl. Oil)
	Food	Non-Food			
1950-52	128	181	77	112	124
1960-62	98	120	77	105	96
1967-69	100	100	100	100	100
1970-72	98	90	86	107	93
1973	111	131	91	139	110
1974	135	123	116	440	133
1975	116	106	100	440	111
1980	98	91	101	400	100

for their import prices. Such cost effects from higher primary commodity prices on the developing countries are best demonstrated by oil: the cost of their oil imports was some $4 billion in 1970-72 on the average, rose to about $5.5 billion in 1973 and to about $15 billion in 1974. Similarly, the costs of their imported foodgrains climbed from $3.2 billion on average in 1970-72 to about $8 billion in 1973 and 1974. And commodities like sugar, timber and cotton are important imports of developing countries as well.

The estimates in Table 4 show, first of all, the deterioration of the commodity terms of trade of all non-oil developing countries which occurred in 1972 as compared to the base period 1967-69. This base period was in no way exceptional: average terms of trade in those years

Table 4. Import and Export Prices and the Terms of Trade by Country Groups, 1972–80 (1967–69 = 100)

	1972	1973	1974	1975	1980	Adjusted Export Growth Rate (%) (1972–80)[a]	GDP Growth Rate (%) (1973–80)
Mineral exporters (excluding net oil exporters)							
Export price	105	155	174	188	296		
Import price	118	152	197	209	296	8.1	6.1
Terms of trade	89	102	88	90	100		
Higher income countries (above $200 per capita)							
Export price	114	154	177	190	276		
Import price	115	149	196	207	293	7.0	5.9
Terms of trade	99	103	90	92	94		
Mediterranean							
Export price	121	156	175	188	273		
Import price	121	155	203	215	303	7.9	7.0
Terms of trade	100	101	86	87	90		
Latin America							
Export price	123	161	189	202	297		
Import price	112	145	188	200	284	7.0	5.4
Terms of trade	110	111	101	101	105		
East Asia							
Export price	94	138	159	171	240		
Import price	112	147	200	211	299	5.9	6.0
Terms of trade	84	94	80	81	80		
West Africa							
Export price	106	162	176	189	292		
Import price	114	147	194	206	292	6.5	6.4
Terms of trade	93	110	91	92	100		
Lower income countries (below $200 per capita)							
Export price	113	141	160	169	232		
Import price	115	148	201	213	302	2.1	2.8
Terms of trade	98	95	80	79	77		
East and central Africa							
Export price	116	155	182	191	269		
Import price	121	152	204	217	309	4.0	3.7
Terms of trade	96	102	89	88	87		
South Asia							
Export price	111	135	149	159	215		
Import price	113	147	200	211	299	1.2	2.7
Terms of trade	98	92	75	75	72		
Total non-oil developing countries							
Export price	113	152	174	187	273		
Import price	115	149	196	208	295	6.5	5.15
Terms of trade	98	102	89	90	93		

Source: International Bank for Reconstruction and Development, estimates and projections.
Note: The average terms of trade for the 1967–69 base period are about equal to those for the entire decade of the 1960s.
[a] At constant prices and annual rates, adjusted for terms of trade changes.

were about equal to those measured for the entire decade of the 'sixties. Countries with a substantial component of manufactured goods in their exports did better on the average through 1972. The improvement of the terms of trade for all non-oil developing countries in 1973 amounted to some 4 percent, mainly in countries dependent on exports of primary commodities.

The situation in 1974 is one of a major deterioration of the terms of trade. Although export prices of manufactured products and primary commodities rose further (the estimated average export price increase is 14 percent, following on an increase in 1973 of 34 percent), this is sufficient to offset the unprecedented increase of import prices which is estimated at 32 percent. The oil price is a major element in this increase (on the order of 18 percentage points), followed by manufactures (contributing 10-11 points) and the remainder is the "backlash" of the increases of primary commodity prices on the developing countries themselves. The average deterioration of the terms of trade for the non-oil developing countries thus amounts to 14 percent; this deterioration is shared about equally among all of these countries. But this deterioration follows a significant earlier improvement in 1973 for those at the higher income level with sizeable exports of manufactured goods; for the lower income countries, by contrast, it comes on top of a small deterioration in the preceding year.

The projections for 1975 foresee little or not change in this situation: both import and export prices are expected to move up by about 13 percent, in line with projected inflation in the industrial countries.

For the remainder of the decade a slight recovery is foreseen on average. By 1980, the terms of trade will still be about 8 percent below the level of the sixties.

The lower income countries are expected to experience a continuing deterioration of their terms of trade beyond the colossal decline of more than 20 percent between 1967-69 and 1974. Although the losses of the East and Central African countries are severe, the outlook for the South Asian countries is far worse. As the South Asian countries comprise about 45 percent of the total population of the developing countries, this is clearly the most significant issue for international economic policies in the years ahead.

The penultimate column of Table 4 presents projected growth rates, in volume terms, of exports corrected for changes in the terms of trade, representing the annual rate at which the purchasing power of each country group's export earnings are expected to increase. The average of 6.5 percent for all non-oil developing countries is unequally divided over country groups; not only do the movements of the terms of trade differ as discussed above, but volume growth of exports also varies among them. The mineral exporters and higher income countries appear capable of sustaining growth rates in their capacity to import from own export earnings of 7 to 8 percent per year, while the growth of import capacity is projected at only 2 percent per annum for the lower income countries. The projected rate of increase for South Asian countries is only 1 percent.

Bargaining strength of the developing countries must be viewed against this background, which suggests that one major objective would be not to raise prices above recent levels but to avoid the terms of trade decline now projected and/or provide greater long term price stability. The next section discusses the market aspects of individual commodities to assess prospects for supply management and price action.

Market Aspects of Primary Commodities

The concerted action to increase the price of oil in December 1973 is generally seen as an example of actions which may lie ahead in other areas. It is therefore essential to understand what lessons the oil price increase actually proffers.

The magnitude of the oil price increase is by now well understood. From an f.o.b. price of $1.90 per barrel in 1972, the petroleum price rose to $2.70 in 1973 and to an average price of about $9.78 in 1974.[8/] Revenues of the major oil exporting countries rose from about $16.6 billion in 1970, to $24 billion in 1973 and to an estimated $110 billion in 1974. Revenues in 1975 are expected to be on the order of $122 billion. These changes involve dramatic shifts in international trade and payments patterns and remove, at least for the present, any foreign exchange constraint on development for most of the OPEC countries. Moreover, the price increases were associated with an embargo on deliveries to selected countries so that fear of oil shortages became inextricably linked with concern about the very high costs of oil.

The extent to which the experience of OPEC is reproducible in other commodities must be evaluated in terms of the structure and characteristics of supply as well as of demand, and also in terms of essentiality.

The Case of Oil

In response to rapidly growing demand, production of oil grew at an average annual rate of over 7 percent between 1955 and 1973. Demand for oil and for imported oil grew rapidly because of the growth of income and industry and because the declining relative price increased the share of oil in the energy market. The rapid increase of U.S. import demand for Middle Eastern oil in recent years constituted an important factor behind the oil price developments. Limited possibilities in the medium run to substitute for oil are reflected in low demand elasticities, although even in the short run these cannot be entirely neglected: higher prices do reduce consumption. The low demand elasticities reflect the fact that consumption patterns are built into the capital structure (e.g., transportation network, power supply facilities and the stock of cars by sizes) and will take time to alter.

The supply of oil in international trade is concentrated in a few countries. In 1973, an embargo and a large price increase met the political objectives of one group of oil producers, while the price increase was supportive of the economic objectives of another. This coincidence of objectives made supply management feasible. Further, supply restrictions can be implemented at little or no cost as oil can be retained in the ground and the employment impact of production cutbacks is negligible. An expansion of supplies from other sources is only possible in the medium to long run and requires considerable investments -- in terms of capital and technology research -- with considerable gestation periods. This puts a horizon of some 7-10 years on the capacity of the oil suppliers to maintain prices which contain a major element of near-monopoly profit.

In terms of essentiality there is hardly an internationally traded commodity which can rival oil, particularly in the medium and short term. Imported oil is a significant element of virtually each country's energy supply and also constitutes an increasingly important feedstock for manufactured products like fibers, rubber and plastics, fertilizers and pesticides.

The changes in market organization were also significant. In the 'fifties and early 'sixties the oil sector was dominated by a small number of vertically integrated multinational companies which were independent for their supplies from any individual producing country. The increasing number of smaller and less-integrated companies eroded the position of the major companies particularly during the 'sixties, increasingly competing with the major companies for crude oil. This weakened the companies' bargaining power on price to the point where producing countries decided that there was no need to even negotiate any more.

This set of characteristics of oil in the world economy needs to be compared with those of other primary commodities, if conclusions are to be drawn with respect to changes in the bargaining position of other developing countries dependent on exports of other primary commodities. The principal "lesson" of the oil price is that the lesson is not clear; the experience is far from completed. It is too early to assess the benefits from the viewpoint of the producers, quite aside from how they may be affected by the economic problems in the industrialized countries and the strains which are evident in the international financial system. It is clear that it is easy to maximize earnings in the short term; but to achieve the same objective over an extended period is considerably more complicated and the outcome is still uncertain.

Over the longer term, the principal conclusion likely to be drawn from the experience with oil is that prices can be raised to the level at which it becomes economic to invest heavily in substitutes. Both the current and the expected price of oil, combined with considerations regarding security of supply, will determine to what marginal cost levels the oil-importing countries are willing to go in the production of substitutes. The higher oil prices, and the greater the certainty that they will remain at high levels, the larger the investments for substitute energy supplies and the greater the risk that the substitution will become irreversible as new output will be protected against competition of imported energy (including oil) at lower prices.

The producer, or group of producers, of the commodity for which the price is raised faces risks on the marketing side and also on the side of the imported goods and services. The risks assumed on the side of the commodity itself depend on the state of technology of substitutes, the amounts of capital and other factors of production required to produce these substitutes in significant quantities, the demand elasticity (the essentiality) for the commodity and the time horizon associated with the investment and production of substitutes combined with the discount applied to future earnings. On the import side these same countries need to take account of their own degree of dependence on imports of goods and services, and the scope which essential imports may provide to their trading partners for price increases which may negate the benefits of their own action. Failure to read any of these factors right can be expensive; success can be rewarding.

Supply Position of Other Primary Commodities

In terms of value in world trade there is no single commodity which can rival crude oil. The average value of exports of primary commodities from the developing countries was $50 billion in 1970-72; of this total almost $20 billion consisted of crude oil. The exports of only five other primary products -- coffee, copper, sugar, cotton, and timber -- each exceeded $1.0 billion on average in those years, and constituted together an export value of just over $10 billion (see Table 5). Another $10 billion of developing countries' exports consisted of some 30 primary commodities; the remaining $10 billion was composed of a very large number of minor items, none of which had an export value of more than $100 million. The gap between the importance of oil and other primary commodities has, of course, widened by the recent price developments; at estimated 1974 prices and trade volumes oil has increased its share in primary commodity exports from 40 percent in 1970-72 to about 60 percent in 1974, at the expense of the share of all other primary products.

Commodities for which developing countries control a share of internationally traded supplies comparable to the case of oil (i.e. more than 70 percent of world trade) are numerous but for most the value of trade is small so that the effort to control supplies, even if effective, would yield rather little benefit. Not that these commodities, however small in world trade, should be neglected by producing countries, particularly

Table 5

EXPORTS OF SELECTED PRIMARY COMMODITIES FROM DEVELOPING COUNTRIES, 1955 AND 1960-1972
(millions US dollars f.o.b.)

	1955	1960	1961	1962	1963	1964	1965	1966	1967	1968	1969	1970	1971	1972
I. FOOD	8,670	9,170	9,140	9,590	10,640	11,530	11,680	11,750	11,690	12,210	12,750	14,570	14,730	17,310
CEREALS	848	775	703	883	980	1,228	1,438	1,333	1,201	1,114	1,121	1,062	1,071	899
Maize	69	202	144	187	229	278	336	387	437	373	403	462	585	330
Rice	475	406	484	485	600	645	666	636	604	584	550	456	414	431
Wheat	304	167	75	211	151	305	436	310	160	157	168	144	72	138
FATS & OILS	383	377	391	406	431	451	490	471	416	483	459	590	644	748
OILSEEDS, CAKE & MEAL	688	852	800	865	990	968	1,019	1,011	851	950	851	893	907	962
MEAT	132	188	177	212	302	357	365	379	334	328	462	556	584	900
Beef	101	173	166	200	288	348	344	353	309	309	441	531	566	886
FISH & FISH PRODUCTS	115	227	268	364	366	431	447	535	543	607	658	858	966	997
Fish Meal	11	47	63	116	122	172	176	225	200	237	239	334	331	287
BANANAS	288	297	312	296	305	338	381	452	461	474	488	481	513	564
ORANGES	84	122	126	152	194	171	174	173	197	212	231	213	285	292
COCOA	572	535	480	465	501	514	494	455	587	634	724	852	712	713
COFFEE	2,220	1,853	1,792	1,819	1,953	2,313	2,153	2,305	2,168	2,454	2,366	2,931	2,611	3,022
TEA	569	569	585	587	602	591	595	540	571	539	458	584	569	617
SUGAR	1,014	1,223	1,330	1,196	1,651	1,702	1,438	1,376	1,478	1,485	1,514	1,855	2,021	2,287
ALL OTHER	1,757	2,152	2,176	2,345	2,365	2,466	2,686	2,720	2,883	2,930	3,418	3,695	3,847	5,309
II. AGRICULTURAL NON-FOOD	4,860	5,000	4,710	4,590	4,720	4,590	4,780	4,900	4,510	4,770	5,485	5,500	5,470	6,610
FIBERS	1,963	1,687	1,768	1,759	1,978	1,863	1,974	1,970	1,759	1,806	1,945	1,921	1,885	1,957
Cotton	1,363	1,091	1,098	1,137	1,301	1,238	1,334	1,315	1,192	1,293	1,457	1,449	1,489	1,569
Jute	388	172	220	195	180	162	243	266	249	203	198	207	162	143
Wool	308	266	313	280	286	263	261	270	226	229	205	177	156	161
Hard Fibers	104	158	137	147	211	200	136	119	92	81	85	88	78	84
RUBBER	1,566	1,709	1,147	1,095	1,028	1,008	1,009	1,013	831	860	1,218	1,105	940	904
TIMBER	379	444	457	473	557	586	646	768	766	936	1,073	1,111	1,198	1,173
ALL OTHER	952	1,160	1,338	1,263	1,157	1,133	1,151	1,149	1,154	1,168	1,249	1,363	1,447	2,576
III. METALS & MINERALS	2,391	2,978	3,010	2,997	3,195	3,835	4,265	4,860	4,940	5,490	6,385	7,310	6,230	6,710
NON-FERROUS	1,541	1,578	1,685	1,754	1,686	1,967	2,509	3,023	3,079	3,368	4,149	4,359	3,590	3,754
Copper	918	960	990	1,055	1,031	1,104	1,396	1,906	1,921	2,177	2,746	2,875	2,162	2,244
Bauxite	55	108	118	131	120	141	168	177	172	175	201	201	210	194
Alumina	13	57	80	82	58	67	105	121	161	164	207	231	239	289
Aluminum	-	22	20	24	33	33	27	37	52	88	113	109	98	113
Lead	164	96	108	93	98	115	137	123	105	104	131	112	98	102
Tin	311	256	291	297	267	377	546	524	540	535	598	657	619	640
Zinc	80	79	78	72	79	130	130	135	128	125	153	174	164	172
IRON ORE & MANGANESE ORE	307	634	632	625	642	804	869	888	851	896	983	1,086	1,124	1,112
PHOSPHATE ROCK	117	141	145	149	169	200	201	218	216	226	220	216	219	252
ALL OTHER	426	625	548	469	698	864	686	731	794	1,000	1,033	1,649	1,297	1,592
IV. FUELS	5,990	7,650	8,100	8,870	9,480	10,630	11,310	12,040	13,260	15,010	16,210	18,100	23,030	27,010
PETROLEUM (CRUDE & PROD.)	5,591	7,128	7,482	8,062	8,723	9,875	10,458	11,131	12,454	13,797	15,050	15,639	20,525	24,205
ALL OTHER	399	522	618	808	757	755	852	909	806	1,213	1,160	2,461	2,505	2,805
TOTAL (I. - IV.)	21,911	24,798	24,960	26,047	28,035	30,585	32,035	33,550	34,400	37,480	40,830	45,480	49,460	57,640

NOTE: Category I, Food, consists of SITC codes 0 (food and live animals), 1 (beverages and tobacco), 22 (oilseeds, oil nuts and oil kernels), and 4 (animal and vegetable oils and fats). Category II, Agricultural Non-food, consists of SITC 2 (crude materials, inedible, except fuels). Category III, Metals and Minerals, consists of SITC codes 27 (crude fertilizers and crude minerals, excluding coal, petroleum and precious stones), 28 (metalliferous ores and metal scrap), 67 (iron and steel), and 68 (non-ferrous metals). Category IV, Fuels, is SITC 3 (mineral fuels, lubricants and related materials).

Source: See Table 1.

November, 1974

in cases where producing countries are only few. For example, the case of pepper, with a value of $78 million[9] in world trade in 1970-72, suggests a possibility for producers' action as only two producers (India and Indonesia) dominate the world market. However, often the administrative and organizational problems are not proportional to the value of trade and a multitude of arrangements for minor commodities is therefore rather unlikely.

There are only 12 commodities for which developing countries' exports exceeded $500 million in the 1970-72 period, accounting for 0.8 percent of their exports; of these the developing countries controlled a major part of total supplies for only seven. Their main characteristics are shown in Table 6.

The Seven Major Export Commodities

The seven products for which developing countries have an equal or larger share in world trade than they have in the case of oil have other supply characteristics which make those commodities less susceptible to supply management. If only control of supplies and the capacity to store the product are considered, the best prospects for supply management are in rubber and tin. The poorest long-term prospects seem to exist for the tropical beverages.

Rubber: Two countries (Malaysia and Indonesia) control more than 70 percent of export trade in rubber; storage is possible at acceptable costs and it takes some 5-6 years for rubber trees to mature. But the existing production capacity for synthetic rubber, which already supplies 68 percent of the market, would make supply management very difficult.

TABLE 6

DEVELOPING COUNTRIES' POSITION IN TRADE OF 12 MAJOR PRIMARY COMMODITIES (AVERAGES FOR 1970-72)

	1970-72 LDC Share in world trade 1/ (in %)	Number of largest LDC suppliers (above 70% of total trade)	World trade as % of world production 1970-72	Gestation period for new investments (years)	Synthetic substitutes as % of total supplies	Storage possibilities
70-100%						
Coffee	97	10	71.5	6.7	..	limited
Sugar	71	30	27.2	3-4	negl.	"
Rubber	98	2	90.7	5-6	68	yes
Cocoa	100	4	78.0	6-7	..	limited
Tea	83	6	58.3	5-7	..	"
Tin	86	4	87.7	2-4	..	yes
Bananas	92	9	20.3	1-2	..	no
50-70%						
Copper	54		58.1	4-5	..	yes
Cotton	57		32.3	1-2	54	yes
Less than 50%						
Timber	26		8.3 2/	..	..	yes
Iron Ore	39		41.4	4-5		yes
Beef	30		5.2	2-3		limited

1/ Gross Exports

2/ Based on 1970-71 Trade Data

A managed price would be reduced in the short run by better capacity utilization in the synthetic rubber industry, which is low at present. Over subsequent years the price would come under pressure from additional manufacturing capacity leading to a larger market share for synthetics. Once synthetics capacity is installed, it can compete at prices which only cover variable production costs and thus maintain its market share.

Tin: Four producers (Malaysia, Thailand, Indonesia and Bolivia) control more than 70 percent of international trade and world trade is 88 percent of world output. Storage is expensive, as is demonstrated by the low costs of operation of the international tin buffer stock. In this case two factors obstruct supply management: the size of the U.S. stockpile approximately equals a year's world consumption (211,000 tons vs. 250,000 tons), and there is competition from aluminum, tin-free steel (TFS) and plastics in some of the main end-uses of tin. In the absence of a U.S. stockpile of tin, supply management could be expected to be effective if undertaken jointly with supply controls for bauxite/aluminum. As both Australia and the USSR are among the top ten producers of both bauxite and tin, concerted action may not be easy.

Tropical Beverages: These three tropical beverages (coffee, tea and cocoa) can be considered as a group, within which substitution can be of significance.[10/] For each of these products a small number of suppliers control a major part of world output and most countries dominating the supply of one of these commodities are also important producers of at least one of the other two products. Gestation periods of new investments in the three crops is 5-7 years; storage is, however, costly and losses can be substantial, particularly in terms of quality.

Past efforts at international agreements for these three commodities, with the objective of managing supplies to stabilize prices, have proved difficult. No international tea agreement has been feasible because of disagreements between traditional and new producers about market shares. In the case of cocoa, an agreement became effective in 1973 after 14 years of negotiation, but no action has been necessary as the market price was (and probably will remain for some more years) considerably above the agreed intervention points; no buffer stock exists as yet. For a number of years, the Coffee Agreement has had the appearance of a successful operation but this was largely achieved through the self-imposed restraint of Brazil which held large stocks and also reduced its share in world production and exports considerably over time. Thus, a large number of small producers, many of them newcomers to the coffee market, could expand their exports while prices gradually rose as a result of Brazil's coffee policies.

These difficulties and problems would be compounded if a multi-commodity approach were attempted, and if the objective were not only to stabilize product prices but to raise them beyond present market levels since this would require an even more rigorous control of supplies.

Sugar: The market situation is more complex and negotiated prices have governed a portion of exports. Although developing countries supply a major share of world trade, the bulk of sugar production in the world is for domestic consumption. The share in world trade is therefore a poor indicator of market strength; a better measure is the share of developing countries' exports in world sugar consumption, which amounted to only 21 percent on average in 1970-72. In addition, this volume of exports is shared by no less than thirty countries and for most of them sugar exports constitute only a minor element in their total export earnings.

Bananas: Efforts at joint action on supply by banana producers are in process. The supply characteristics of the market, in terms of the possibility of joint action by producers, are mixed. Bananas have been in chronic oversupply which has permitted the few companies which dominate the collection, transport and distribution to be highly selective in their procurement from sources other than their own plantations. In terms of negotiating higher prices, the bargaining strength of the major producing countries is, however, enhanced by the fact that they face only a small number of companies, provided that these countries can form a united front in order to forestall attempts by the companies to reduce their offtake from any one country at a time. Because of the chronic oversupply, effective action must include a rigorous quota system and limitations on production. The reluctance of Ecuador, one of the major producers, to join a producers group does not augur well. Ecuador had a strong interest in joint action during the 1960's when bananas accounted for more than half of the country's export earnings but the share of bananas in exports declined to about 40 percent in the early 1970's. The recent rapid increase of the oil exports earnings has made Ecuador much less interested in the banana market though it has enhanced its financial capacity to sustain supply management action.

The potential of control over supply is made more difficult by the impossibility of storing bananas. Excess production at a given price is lost for consumption and this loss must be added to the cost of the output which is sold. The cost of covering such losses due to curtailment of sales could offset the benefits from higher prices in net terms even if demand is price inelastic, unless total production is brought into line with demand.

The seven major commodities discussed above, except tin, have common characteristics in terms of employment. All of them are labor intensive agricultural products and supply management has therefore significant consequences for employment. This is further emphasized by the fact that these crops usually are grown in particular regions of the producing countries where the production of these commodities is a major element of regional incomes. A country participating in supply management in the world market may therefore gain in overall terms as foreign exchange earnings are increased above the level attainable without supply management, but higher unemployment and loss of income in the producing regions may, without offsetting actions, skew the benefits within the country. Investments in alternative land use and substituting employment opportunities will be required; those need to be planned carefully and in time in order to avoid undesirable social consequences.

Five Additional Important Commodities

For the five other major commodities of which developing countries' exports exceeded $500 million in 1970-72, their share of world trade is less than 60%. In the case of cotton, beef and iron ore trade is relatively small compared to world production so that the points raised above in respect to sugar apply here as well. Timber and copper prospects are more promising.

Timber is a more promising item as seen from the supply side as the share of developing countries in world timber trade does not reflect their true market strength. Tropical hardwoods constitute a market of their own within the timber trade because of their particular end-uses and because supply is concentrated in a few Southeast Asian and West African countries. There can be little doubt as regards the possibilities of increasing prices through the curtailment of supplies; in fact, the main producers in Southeast Asia have already announced their intention to place strict limits on the production of tropical hardwood logs dictated largely by the need to avoid depletion of the timber resources. The world's resources of these types of timber are not large and take considerable time (e.g. 50-70 years for teak) and good management to regenerate. An already existing scarcity will consequently become more severe. The development of substitutes and production of end-products which use less hardwood can be expected to accelerate.

The international copper market is one in which suppliers could obtain limited benefits. The expansion of copper mining and processing facilities is costly and requires long lead times. Four major producers (Zambia, Zaire, Chile, and Peru), account for 36 percent of world production, (excluding centrally planned economies) and for 78 percent of the exports of net exporting countries. They could therefore effectively influence the supply in the world market if they decided to do so and if the developed country producers (U.S. and Canada) did not seek to undermine the effort.

The production of copper from scrap in the consuming countries is, however, of some importance and reduces the market strength of the four major producers, particularly when higher copper prices would stimulate recovery from scrap. The recent announcement by the four major exporters of a ten percent reduction in copper shipments which is to be implemented over the course of the next few months indicates their willingness to act in a joint fashion. However, it should be noted that this reduction is thought necessary to keep the copper price from falling further -- it was $1.26 per pound in the second quarter of 1974 and $0.63 in October -- in the course of 1975; a much larger reduction in shipments accompanied by reductions in output would be needed to restore the 1972-73 average price.

World copper resources are large and a substantial part is in the developed countries. Exploitation of those resources would become attractive if the present major exporters managed to restore a high price of copper. In the long run, when new production entered the market, reducing the share of the present exporters in the world market the price would fall again. Benefits of such price management would therefore accrue only for a limited time. But benefits obtained in the earlier years could be used to finance accelerated diversification of the producers' economies; that would generate new exports to compensate in later years for the reduced revenues from copper. Depending on the size of the price increase and the period over which that price can be maintained (and those two are not unrelated), the net benefits of such action may be judged adequate to undertake supply management. However, possibilities for diversification differ widely among participating countries; hence net estimated benefits differ as well. The risks of such action may then be judged differently by each participating country, and its willingness to join for a sustained period of time would depend on its share of benefits.

Other Commodities

Other commodities for which developing countries supply a significant amount in world trade, but which are less important in terms of value are hard fibers (abaca, sisal and jute), bauxite, manganese ore and phosphate rock. Average annual exports of those products from developing countries in 1970-72 were between $100-200 million.

Fibers do not offer any prospects as their existing markets already are threatened severely by synthetic substitutes and their market shares have already eroded considerably. The supply characteristics of the three minerals are similar to copper.

Supply Policies for Major Minerals

Supply management for the major minerals and metals will benefit only a small number of countries; since there are relatively few major producers of each commodity which depend principally on these commodities for their export earnings. (See Table 7).

Table 8 compares the rates of growth in production of the major minerals with those of oil since 1955. Even where technological substitution is feasible, producers of a commodity in consistently strong demand obviously are in a better negotiating position than producers of other commodities. Of the major minerals shown in Table 9, production of bauxite and related products and phosphate rock grew at about the same rate as oil output; lead, tin, zinc and manganese production grew rather slowly. Iron ore and copper output grew at intermediate rates.

The four major minerals (bauxite, copper, manganese, tin) accounted on the average in 1970-72 for less than 6 percent of total exports of the developing countries; there are in total only 10 countries for which

TABLE 7

MAJOR MINERALS EXPORTERS IN VALUE OF WORLD TRADE
(1970-72 average; percentages)

	BAUXITE		COPPER		MANGANESE		TIN	
	country's exports	world trade	country's exports	world trade	country's exports	world trade	country's exports	world trade
Jamaica	25	32	-	-	-	-	-	-
Surinam	30	17	-	-	-	-	-	-
Guyana	13	7	-	-	-	-	-	-
Chile	-	-	72	16	-	-	-	1
Zambia	-	-	94	17	-	-	-	-
Zaire	-	-	68	10	1	3	3	3
Peru	-	-	22	5	-	-	-	-
Gabon	-	-	-	-	21	20	-	-
Malaysia	-	2	-	-	-	-	18	43
Bolivia	-		2	-	-	-	52	14
Other LDCs	a/(12)	15	(5)	6	(4)	33	(17)	24
TOTAL	0.35	73	4.21	54	0.19	56	1.11	85
Value of world trade (average 1970-72) in U.S. $ millions		278		4,495		194		747

a/ Figures in brackets indicate the number of other exporting countries.

Source: IBRD: Commodity Trade and Price Trends, Report No. EC-166/74, August 1974.

TABLE 8

WORLD /a PRODUCTION OF SELECTED MINERALS, 1955-73

	1955	1960	1965	1970	1972	1973/b	Average Annual Growth Rates
CRUDE PETROLEUM ('000 bbls.)	5,016,972	6,466,650	9,070,507	14,899,300	16,406,750	17,766,375	7.3
METALS AND MINERALS ('000 M.Tons)							
Non-Ferrous							
Copper (copper cnt.)	3,112	4,242	5,066	6,374	7,049	7,519	5.0
Bauxite	17,760	27,620	37,292	59,484	68,860	73,134	8.2
Alumina	6,200	9,300	13,600	21,095	23,440	26,183	8.3
Aluminum	3,105	4,528	6,586	10,207	11,513	12,708	8.1
Lead	2,178	2,376	2,750	3,438	3,452	3,477	2.6
Tin (tin cnt.)	194	189	191	219	232	233	0.8
Zinc (gross wt.)	2,967	3,351	4,229	5,333	5,477	5,670	3.7
Iron Ore (Fe cnt.)	174,500	222,100	325,000	417,700	417,700	468,000	5.6
Maganese Ore (Mn cnt.)	4,709	5,524	6,797	7,900	9,014	9,699	4.1
Phosphate Rock	28,591	39,445	60,375	81,074	88,819	98,776	7.1

a/ Crude Petroleum data exclude Centrally Planned economies.

b/ Preliminary

SOURCES: U.S. Bureau of Mines; BP, Statistical Review of the Oil Industry; U.N., Statistical Yearbook, 1962; Metallgesellschaft, Metal Statistics (various issues); IBRD, Economic Analysis of Maganese Market, Working Paper No. 104; UNCTAD, Manganese Ore: Problems of Trade Liberalization and Pricing Policy, February 27, 1974; UNCTAD, Phosphates: Problems of Trade Liberalization and Pricing Policy, March 13, 1974.

these commodities are of significance in their exports, while at the same time accounting for a considerable part of world trade. These countries also belong mainly to the group of higher income developing countries, while the prospects for supply management and the maintenance of higher prices for several of the mineral products is feasible, even if for limited periods, it has only limited significance for the developing countries as a group.

In terms of the possible impact on international prices and on export earnings, the significance of these commodities also is much less than that of oil. Average annual exports by developing countries in 1970-72 amounted to $3-4 billion for the four major commodities listed above, whereas oil exports from the developing countries amounted to $20 billion. The price increase between that period and 1974 added $75-80 billion to the revenues of the oil producers, but for the other minerals a similar price increase (which would in any case be an unlikely scenario) would add at most $13 billion to export earnings. For the minerals, the expectation is that prices in constant dollars will remain above the 1967-69 level and there may be possibilities to mitigate a decline from the high 1973-74 price levels either through joint action or through unilateral price leadership.

The action of producers to manage supplies in order to raise prices need not be limited to developing countries, if the objective is only to maintain or to increase prices. Developed country producers may well join in the effort, as evidenced by the recent position taken by Australia and apparently also by Sweden in connection with iron ore and bauxite. The implications of this possibility are of a major importance, particularly in the field of metals and minerals. If Canada would also become involved producers would be in an extremely strong position to set prices of their exports.

The Demand Structure and Essentiality

For each of the major primary commodities discussed above, the main markets are the developed countries. Significant differences between the twelve major commodities relate partly to the geographic distribution of demand and partly to demand response to price.

The geographic distribution of imports is heavily weighted by Western Europe and North America, and in the case of minerals also by Japan (see Table 9). For tea, cotton and sugar the developing countries themselves are major importers; the centrally planned economies are a sizeable market for rubber, cotton and sugar. This would suggest that, because of the "spread" of importing countries, the bargaining position of consumers is relatively weak in the cases of sugar, tea, cotton and rubber. But this conclusion needs to be drawn with caution. The sugar market is dominated by production for domestic use and further characterised (at least through 1974) by three major preferential arrangements. The tea market is dominated by one consumer, the United Kingdom.

The demand for tropical beverages except for tea is heavily concentrated in North America and Western Europe. Because of their traditional character, the income and price elasticities tend to be low in those markets, at least within the historical price ranges for those products. Price increases for these products would therefore likely be of benefit to the producers as these would only partly be offset by a reduction of of the volume of demand. Consuming countries could retaliate with

TABLE 9

SHARES IN VALUE OF WORLD IMPORTS OF MAJOR PRIMARY COMMODITIES, BY AREA
(1970-72 Average; Percentages)

	U.S. & Canada	Western Europe	Japan	Other Developed	Centrally Planned	Developing	Total
Coffee	39	50	2	1	4	4	100
Sugar	29	24	11	1	17	18	100
Rubber	20	32	9	3	27	9	100
Cocoa	24	50	3	2	18	3	100
Tea	11	41	3	6	8	31	100
Tin	29	41	14	-	6	10	100
Bananas	27	47	18	1	2	5	100
Copper	9	64	19	-	3	5	100
Cotton	2	32	18	1	26	21	100
Timber	15	42	25	2	5	11	100
Iron Ore	13	39	36	-	11	1	100
Beef	30	58	2	1	4	5	100

Source: U.N. Trade Statistics.

indirect taxes, as these products are not essential. Consuming countries could attempt to reduce the consumption volume sufficiently to negate any positive effect on the total revenues of exporters. Although this may be difficult to achieve, the possibility of tax retaliation is a risk for the producing countries.

Higher prices could also accelerate the consumption of substitutes, in particular the soft drinks which have rapidly increased their share in the market. The manufacturing of soft drinks permits rapid increases in their supplies and the distribution and consumption of these products is simpler than for the tropical beverages. In countries which have experienced rapid income growth in the last 10-20 years and in which the traditional tropical beverages had only a small initial market position, _per capita_ consumption levels may never equal those of the developed countries. Long-run benefits from price increases are therefore quite uncertain.

More or less the same applies to bananas, although in this case there is a very wide gap between the price per volume unit as compared to other fruits. A significant price increase for this cheapest type of fruit may therefore be more readily accepted by consuming countries. Taxation of domestic consumption would in any case be difficult, as it is for all perishable commodities.

The possibilities in the case of sugar are minimal, as was stated before, because of the large sugar production capacity in the consuming countries themselves. The only way for developing countries to increase their earnings from sugar is to press further for trade liberalization as sugar production in many developed countries involves high costs and takes place behind protective shelters. Liberalization would not only reduce such production, but also increase demand which would benefit the developing countries significantly if they could increase their own production sufficiently.

Rubber and tin, although consumed in a large number of countries, face competition from substitutes which suggests that unilateral supply and price management for any extended period is not feasible, although to some extent both these products benefit from the high oil and energy costs embodied in their substitutes, notably synthetic rubbers but also aluminum, plastics and tin-free steel. This raises the floor price for these two commodities but substantial further price increases could not be maintained for any length of time. About the same reasoning applies to cotton in its competition with synthetic fibers, but imported cotton also has to meet the competition of production of cotton in North America. Beef imports have again the same feature, competing with beef production from the developed countries. In addition, non-tariff barriers have become widespread and their removal could be far more beneficial than any attempt to raise prices unilaterally.

This leaves three commodities of a more essential nature from the importers' point of view: copper, tropical timber and iron ore. An interesting feature of all three is the relative importance, or even dominance, of the Japanese and Western European markets which account for some 70-80 percent of world imports of these three commodities, whereas the North American market accounts for only 10-15 percent. This concentration of demand reflects the poor resource base of Japan and Western Europe for these commodities and strengthens therefore the bargaining position of the producing countries.

Timber prices were stable in real terms throughout the 1960's as Japan and other East Asian countries expanded their demand for tropical hardwood, and import demand in the United States rose. A high income elasticity and a relatively small response to price in the consuming countries have led to a situation where the few producing countries are rapidly gaining bargaining strength and are now ready to use this to their advantage. Although wood substitutes are already widely used, these have not so far made wood less desirable for quality products. Takeuchi's market analysis[11/] for tropical hardwood from the main producing area, Southeast Asia, suggests that producers will remain in a strong market position well into the 1980's.

Copper offers, next to timber, the most interesting case as Western Europe is by far the most important market, followed by Japan. Both have very limited potentials for domestic production. America is self-sufficient. Supply restrictions and higher prices by the major developing copper producers would therefore by-pass the United States and Canada, which have undeveloped copper resources, but be felt by Western Europe and Japan which depend on imports for more than 90 percent of their copper consumption.

Iron ore is in a less strong position; although supply has shifted in recent years to a number of new producers like India and Brazil, known resources are very large and are now brought into production in developed countries as well, notably in Australia. Price action would have to be joined by Australia to be effective (as in the case of bauxite) but even then success is in doubt as in the past, demand for ore has not grown at the same rate as steel production due to increasing use of scrap. Increased input flexibility of steel mills could weaken the position of the suppliers of ore. Some strength derives, however, from the increased vertical integration through use of pelletization processes and special carriers, and from the dominance of trade under long-term contracts.

Secondary Effects of Price Action

Price action for export commodities must, of course, be seen in context. While it increases the revenue to the producing countries, it reduces the income of others. Inflation may act differentially on the import basket so that part of the price gains are eroded through terms of trade losses, but one need not complicate the general analysis by such an assumption. Aside from inflation, the increased cost of raw materials will be reflected in the finished products which the primary exporters import. The net gain to the exporters will thus depend on the share of the commodity subject to price action in total exports versus the cost increase in the manufactured goods component of imports. Since, in general, developing countries have trade deficits, and since a country exporting a single primary commodity obviously imports finished products which may reflect price increases in several primary products, the offset could be substantial. Although the bulk of the reduction in real income will fall on the principal consumers, part of both the direct and the indirect effects will fall on the developing countries.

This, of course, has also been the experience in the case of oil. While the current account balance of the industrialized countries deteriorated by $53 billion as a result of the oil price increase, the import bill of the developing countries rose by about $15 billion in 1974. The indirect effects, which would be diffused over the intermediate and final products in the import basket, are more difficult to quantify but could easily be on the order of an additional $3-4 billion.

These factors do not militate against an increase in price where this is feasible, since primary commodity producers cannot be expected to take a more global view of the pricing of their product than producers of other commodities. They can argue legitimately that other mechanisms should be found to support and accelerate the growth in real income in other developing countries. Nonetheless, the results need to be borne in mind since they suggest that the potential bargaining capacity of one group of countries cannot be equated with improvement of the welfare of the developing countries in general.

The Broader Objectives of Bargaining

The conclusion of the preceding analysis is that only few primary commodities show characteristics which would support the possibility of raising prices through arrangements among producers. The countries which produce the commodities for which some possibilities may exist belong to country groups which already have made significant development progress and can pursue their growth objectives further in the coming years, *albeit* at rates which may be somewhat reduced as compared to the last five years. The countries experiencing the greatest difficulties also have the worst chances of development and the least capability of mobilizing their own resources for growth; they are also the weakest in the international markets where they trade. As these countries have the major part of the population of the developing world, they represent the major problem of economic development. Changes in bargaining strength between countries have not made that task any easier: quite to the contrary, their problems have been exacerbated by the improved bargaining power of other developing countries and may be so again in future. When discussing the effect of improved bargaining strength on the totality of the developing world, the skewedness of the distribution of likely benefits and costs must therefore be of prime concern to the participants in any bargaining process.

On the assumption that prices for some commodities can be raised through supply management, we next address the question whether this can be translated into bargaining power related to broader economic or political objectives. It is on this issue particularly that the distinction between oil and other commodities is important, lest inappropriate lessons are drawn from the recent experience with oil.

Commodity prices are a poor bargaining tool for this process since successful negotiations imply a commitment to a production policy which may not be profitable over the long-term and presumably few countries would be prepared to sell commodities at prices below those which they could otherwise obtain. At most, the producers could promise not to restrict output through administrative measures but even then, they would be reluctant to forego the possibilities of price stabilization agreements which might involve production controls. Thus, as far as price is concerned the bargaining of the producers must rest on a threat to raise prices.

As discussed earlier, the level to which prices can be raised is limited by a number of factors, including the share of production controlled, demand response, immediate substitution possibilities, medium-term technological alternatives and the possibility of major investments in alternatives. Most minerals, aside from oil, do not represent a major component in consumer prices so that, while price increases for the raw materials obviously would be reflected in increased final product prices, they are likely to have a much more muted effect than in the case of oil. For other minerals, even if prices of all were to be doubled, it might yield an additional $6-7 billion which should be compared to the $80 billion in increased oil revenues. The threat of such an increase is probably not sufficient to obtain action on non-price related matters, particularly since the period over which such prices could be maintained is expected to be relatively short. Moreover, it is not reasonable to assume that some of the industrialized producers would participate in an action not related to commodity prices so that the percentage of production affected might well be less for a non-commodity related action.

In terms of non-commodity related objectives, access is more important than price, though there are obvious links between the two, short of an embargo. In the case of oil, the embargo was in support of a political objective shared by several OPEC members who also are among the principal producers. In that case, the geographic concentration of oil reserves was a crucial factor since it coincided with a major regional political issue. Other minerals are not so concentrated so that it is rather difficult to envisage an analogous identity of political purpose. Moreover, the dependence of many of the mineral exporters on current export earnings makes such an action, if not unlikely, at least much more expensive.[12/] Non-political issues, e.g., bargaining about general trade relationships, or the SDR link, or voting rights in international institutions, or aid flows, are unlikely to be made so confrontational as any embargo action inevitably implies.

In assessing bargaining strengths and objectives at any level, it is of course important to assess the strengths of both sides. While economic power is considerably more diffused than in the 1950's, if the issue is drawn between primary commodity producers and other countries, the overwhelming economic strength lies with the latter. At a simplistic level, all activity depends on the availability of raw materials, but the developing countries do not control the latter and, as hardly needs pointing out, raw materials by themselves do not get anyone very far. This is merely another way of saying that the world is increasingly interdependent and in such a world confrontations are not likely to be beneficial for long to any party. As the analysis of the trade data shows quite clearly, the developing countries as a group remain heavily dependent on the industrialized countries for markets, for research and technology, for investment goods and, to an increasing extent, for such strategic requirements as food and fertilizer. If it is legitimate for governments to limit production in order to set the export price for a primary product, it would be equally legitimate for governments to seek to maintain, say, fertilizer export prices; if the price of an LDC primary export is to be made immune from price and exchange rate changes, why not also the price of wheat or for that matter generating equipment. In an inter-dependent world, real national interests must be seen in broad terms and bargaining can only take place in the framework which recognizes the relative needs and the interests of all participants.

Footnotes

1. The authors are, respectively, Director, Development Policy and Director, Economic Analysis and Projections Department, the World Bank. The article represents their personal views only; the views are not necessarily those of the World Bank.

2. Europe - The Way Ahead, OEEC, December, 1952, pp. 26-27.

3. "The Threat from the Third World", Foreign Policy II, Summer 1973.

4. "The Response to the Third World," Foreign Policy, Winter, 1974-75, p. 6.

5. See for example -- "Problems of Raw Materials and Development," TD/B/488, Report of the Secretary General of UNCTAD to the Sixth Special Session of the General Assembly, United Nations, New York, 1974.

6. This notably the case for Commonwealth preferences which were eroded by the two successive rounds of GATT tariff reductions and were finally abolished through the United Kingdom's joining of the European Community.

7. See for example - Lester R. Brown, "The Interdependence of Nations," Overseas Development Council Paper No. 10, Washington, October 1972.

8. Prices for the "marker" crude, Saudi Arabian Light, 34°, f.o.b. Ras Tanura; average f.o.b. export prices of all OPEC countries together are higher because of quality differences and freight differentials.

9. Excluding Singapore (re-exports).

10. Efforts to demonstrate the degree of substitution between these three commodities have, however, been largely unsuccessful.

11. K. Takeuchi: Tropical Hardwood Trade in the Asia-Pacific Region, Occasional Paper No. 17, IBRD, 1974.

12. There is, of course, the possibility of oil exporting countries financing commodity stockpiles to support drastically higher prices while offsetting the domestic impact of reduced production or even to finance imports in support of a prolonged embargo. Quite aside from the question of whether large-scale investing in commodity futures is a prudent use of funds derived from a diminishing resource, this type of major confrontation almost surely would evoke retaliation.

ANNEX TABLE I

MAJOR PRODUCERS OF SELECTED PRIMARY COMMODITIES, 1960-62 AND 1970-72 AVERAGE

	Coffee	Copper	Sugar	Cotton	Timber 2/	Iron Ore	Rubber	Cocoa	Beef	Tea	Tin	Bananas
1960-62 AVERAGE												
	Brazil	U.S.	USSR	U.S.	U.S.	USSR	Malaysia	Ghana	U.S.	India	Malaysia	Brazil
	Colombia	Zambia	Cuba	USSR	China P.R.	U.S.	Indonesia	Nigeria	USSR	Sri Lanka	Bolivia	India
	Angola	Chile	Brazil	China P.R.	Brazil	France		Brazil	Argentina	China P.R.	China P.R.	Ecuador
	Ivory Coast	USSR	India	India	Canada	China P.R.		Ivory Coast	China P.R.		Thailand	Venezuela
	Mexico	Canada	U.S.	Mexico	India	Sweden			France		Indonesia	Honduras
			France		Indonesia	Canada			Germany F.R.		USSR	Thailand
			Germany F.R.		Sweden	India			U.K.			Colombia
			Australia		Nigeria				Australia			Philippines
			Mexico		Japan				Canada			Panama
			Philippines		Finland							
			Other (8) 3/		Other (23) 3/							
Total % Accounted for:	70.4	71.2	70.5	72.5	70.1	72.4	70.0	72.9	70.8	70.3	77.0	69.5
1970-72 AVERAGE												
	Brazil	U.S.	USSR	U.S.	U.S.	USSR	Malaysia	Ghana	U.S.	India	Malaysia	Brazil
	Colombia	USSR	Cuba	USSR	China P.R.	U.S.	Indonesia	Nigeria	USSR	Sri Lanka	Bolivia	India
	Ivory Coast	Chile	Brazil	China P.R.	Brazil	Australia		Brazil	Argentina	China P.R.	China P.R.	Ecuador
	Angola	Zambia	U.S.	India	Canada	France		Ivory Coast	China P.R.	Japan	Thailand	Burundi
	Ethiopia	Canada	India	Pakistan	India	China P.R.			Brazil		Indonesia	Honduras
	Uganda	Zaire	China P.R.	Brazil	Indonesia	Canada			France			Costa Rica
	Mexico		France		Sweden	India			Germany F.R.			Philippines
	Indonesia		Australia		Nigeria	Sweden			Australia			Mexico
	El Salvador		Mexico		Japan				Canada			Thailand
			Germany F.R.		Finland				U.K.			Venezuela
			Other (7) 3/		Other (23) 3/				Other (1) 3/			Other (1) 3/
Total % Accounted for:	72.6	74.2	70.7	72.4	70.5	73.7	68.5	71.8	70.1	70.0	75.5	70.2

Those accounting for about 70% of world production
Based on 1961-62 average
Number of countries in brackets

Source: IBRD. Data bank.

III. DOMESTIC ADJUSTMENTS AND ACCOMMODATIONS TO HIGHER RAW MATERIAL AND ENERGY PRICES

T.L. Sankar

The sudden and unexpected increase in the price of oil and its products and the more gradual increase in the price of several raw materials have created serious imbalances which call for urgent domestic and external adjustments in all developed and developing countries. This paper attempts to briefly survey the extent of the increase in energy and raw material prices insofar as they affect the developing countries, to identify the likely effects of such price increases, and to examine the feasibility of domestic adjustments in these countries. It is difficult to discuss these issues with any degree of specificity because of the variations in the relative affluence (or poverty) of the developing countries, the diversity of their resource endowments, the wide variations in their technological capabilities and the uncertainties regarding future oil and other raw material prices. The discussion is therefore vague and generalized and is meant only to stimulate further examination of these issues.

Nature and Magnitude of the Problem

The examination of the impact of increased prices of oil and raw material must be seen in the context of the increase in the prices of other goods. The increase in international prices became a major issue at the turn of the decade and started - although in a relatively low key - with the inflation in developed countries. "It began before the rise in the prices of petroleum and other primary commodities and it is only partially explained by them."1/ The prices of capital goods and manufactured goods exported by developed countries, which rose less than 6% in the decade of the sixties, have risen by more than 10% annually since 1970. (See Table 1).

TABLE 1

Index of Export Prices of Developed Countries[1]

	1956	1968	1972	1973	1974
1. Index (1968-69 = 100)	94	93	128	154	175
2. % change per year over figures in previous col.	-	0.4	6.9	20.5	14.0

Source: Robert McNamara: Address at the Fund Bank Annual Meeting, IBRD, Washington, Sept. 1974

[1]Note: An index of capital goods and manufactured export prices of major developed countries. The index also reflects changes in exchange rates.

The prices of primary commodities, which constitute the major exports of developing countries, showed remarkable stability until 1971, but started to increase steeply from then on. As shown in Table 2, the price of petroleum, which moved slightly ahead of the prices of other primary commodities, showed a very steep increase in 1973 and further steep increase in January 1974.

An analysis of the effects of such violent price changes on the developing countries would involve the determination of the price changes country by country, the changes in the volume of trade, the changes in the commodity composition of trade and the likely changes in the distribution of trade between the developed and developing countries as well as the distribution among the developing countries. A study[2] of the hypothetical impact of the price changes between 1970 and 1974, applied to

TABLE 2

Index of Export Prices of Primary Commodities
1967-68 = 100

Commodity	Index of prices in 1970	1971	1972	1973	1974
1. Food and Beverages	115	116	131	181	297
2. Agri. Raw Materials	95	92	115	201	213
3. Minerals and Metals	111	102	104	146	175
4. All Commodities (except petroleum)	108	105	119	176	244
5. Petroleum	96	123	139	200	665

Source: IBRD-IDA - Additional External Capital Requirements of Developing Countries, Washington - March '74

the country by country trade data of 1970 (without making allowances for the change in the volume and commodity composition of trade from 1970 to 1974), indicates that the trade balance of the developed countries changes from a deficit of $ *8.4 billion to $ 82.1 billion - i.e., from 3.6 percent to 17.5 percent of their imports. The developing countries, whose trade was in near balance in 1970, show and increase in their trade surplus of $ 60 billion, which is about 37 percent of their exports. But the entire increase in the surplus of the developing countries (and more) accrues to the oil-exporting developing countries. These show an increase in trade surplus from $ 8.6 billion in 1970 to $ 66.8 billion in 1974. The deficit in the trade balance of the non-oil-exporting developing countries** registers an increase from $ 8.5 billion to $ 17.4 billion. (30

* $ in this paper refers to US dollars in current terms.
** Hereafter referred to as the "other developing countries."

percent of their imports in 1974 prices.) Even this order of increased deficit in developing countries is considered to be an underestimate. Studies [3,4] which allow partially for the changes in trade in volume and commodity composition beyond 1970 indicate that the likely trade surplus of the oil-exporting developing countries and the trade deficit in the non-oil exporting developing countries would be somewhat higher.

In order to make the analysis more meaningful, it is necessary to disaggregate the developing countries into sub-groups and analyze the impact of increased energy and raw material prices on each sub-group. Although it is possible to effect such disaggregation in terms of the developing countries' endowments of resources (particularly energy and mineral resources), their relative technological capabilities, or the variations in their GNP, for the purpose of this paper, the developing countries are classified into four categories as follows:

1. Oil exporting developing countries.
2. Other developing countries with higher income: per capita GNP above $ 340 in 1970.
3. Other developing countries with middle level income: per capita GNP between $ 200 and $ 340 in 1970.
4. Other developing countries with lower income: per capita GNP below $ 200 in 1970.

A representative selection of countries in groups 2 to 4 was made and the effects of the price increases on their trade balances were examined. The countries selected and their populations are set out in the Annex to Table 1.

On the basis of the sample chosen, as shown in Table 3, higher income group countries have been the least affected among the other developing countries while the middle level income group countries have been the most affected.

TABLE 3

Increase in Trade Deficit Between 1970 and 1974 of Selected Developing Countries

	In Hundred Million $			
	Trade Deficit in		Increase in trade deficit	Col. 4 as % of Col. 2
	1970	1974 prices of 1970 data		
Category 1	2	3	4	5
Non-oil exporting developing countries with:				
1) Higher income	31.4	55.7	24.3	77.4
2) Middle level income	6.5	16.2	9.7	149.2
3) Lower income	13.9	28.8	14.9	107.2

Source: For Col. 2 and 3, as computed and set out in Annex Tables II & III.

Possible Responses

The increase in trade deficits could be balanced either by inflows of capital resources from other countries, by an increase in the terms of trade, or by adjustments in production, consumption, and trade plans. If the prices of the exports of the "other developing countries" could be increased immediately to levels which would neutralize the increase in the prices of their imports, the problem could be solved at once. If the terms of trade improve gradually in the next few years, the problem of increased prices will become a transient one and can be tackled by domestic adjustments. But if the 1974 terms of trade continue throughout the foreseeable future, domestic adjustments of a more enduring kind must be made. The single most important factor determining the nature of the response of the other developing countries to the increased price of oil and raw materials is their perception of the terms of trade that will face them in the coming years.

Terms of Trade

The major export commodities of the "other developing countries" are minerals and agricultural raw materials, the demand for which is a function of industrial activity in the developed countries, and tropical agricultural consumer products (such as sugar, coffee and cocoa), the demand for which is dependent on the level of consumption in the affluent countries. Among these export items, natural products like rubber and jute, which compete in the international market with petroleum-based chemical substitutes, can be expected to benefit directly from the oil price increase. But prices of other agricultural raw materials and minerals depend on the level of industrial activity in the developed countries, and the new demand generated in the oil exporting countries. All indications are that industrial activity in the developed world will not increase at the rapid rate witnessed in the last decade. The rate of growth of the GNP in the OECD countries, which was 5.8 percent in 1972 and 8.7 percent in 1973, is now anticipated to be only 4.8 percent per year[5/] until 1980 - even this is considered a "relatively high rate estimated to ensure that the energy requirements are not understated." The year 1974 has ended with unmistakable signs of a significant slowing down of industrial activity in the USA, Japan, and most countries of Western Europe. Unless the countries exporting minerals and agricultural raw materials organize collective action, the prices of these commodities may come down. While the increased affluence of the oil exporting countries will increase, the demand for tropical products such as sugar, coffee, and cocoa, the resulting demand increase will not be so high as to affect the prices significantly. In sum, there is no indication, based on purely economic considerations, that the terms of trade of the other developing countries will improve significantly after 1974. In fact, a more detailed analysis of the medium term price trends to 1980[6/] leads to the conclusion that the terms of trade are likely to deteriorate further, as outlined in Table 4.

TABLE 4

Terms of Trade of Developing Countries in 1973 & 1980
(1967-68 = 100)

Countries	Terms of Trade in 1973	Terms of Trade in 1980
1. Oil producing countries	140	350
2. Mineral producers	102	102
3. Other developing countries		
A. with per capita income over $ 200	104	95
B. with per capita income below $ 200	94	77

Source: Robert McNamara: Address at the Fund Bank Annual Meeting, IBRD, Washington - September 1974

Transitional Adjustments

If the terms of trade do not improve for the other developing countries, there will be a net outflow of resources from these countries amounting to at least $ 10 billion, about 3 percent of their Gross Domestic

Product (GDP). This must be financed by reductions in consumption or investment. Such reductions can be avoided temporarily by drawing on reserves, or it can be postponed by the flow of capital from other countries to the developing countries by way of commercial borrowing, foreign investment, or aid.

To what extent will it be possible to effect these temporary adjustments? The "reserves" of the other developing countries amount to $ 32 billion, or about 37 percent of their imports in 1970. Nearly 50 percent of these reserves are in the higher income developing countries, and they alone can draw upon their reserves for any significant amounts. The average borrowing in the bond markets of the world by the developing countries increased in the early seventies and is more than $ 6 billion a year at present. But here again a substantial portion of the borrowing is done by the higher income countries and a few middle income countries. The official Bilateral Development Assistance (BDA) disbursed to the other developing countries during 1969-72 has been on an average of only $ 3.6 billion annually; over 50 percent of this was for the lower income countries. Among the developing countries, the higher income group may be able to step up their withdrawal from reserves and their level of commercial borrowing, but the countries with relatively lower income will have to depend on development assistance. Unless the oil-exporting countries, which have a sudden increase in their surplus, recycle the same to the lower income developing countries through bilateral negotiations or through appropriate intermediaries, the lower income developing countries may not be able to effect even the temporary adjustments.

However, the capital flows by themselves are unlikely to provide a lasting solution to the problem. Higher costs of energy and raw materials will have to be paid for either by an improvement in the terms of trade or by adjustments in production, consumption, and trade. At best, capital flows provide a breathing space during which such adjustments can be made. Ultimately, a reduction in consumption or investment is necessary. Depending on the proportion of these two elements, the time path of present and future consumption will be altered. If the adjustments are financed entirely by reduced consumption, there will be an immediate reduction in consumption; if they are wholly financed by reduction in investment, consumption will be reduced some time in the future. If adjustments are financed by borrowings, consumption levels will be altered and the time profile of the new level of consumption will depend on the extent to which borrowed funds are used for increasing current consumption or investment and the terms of repayment of the loans. Since debts incurred have to be repaid at some date - however distant - borrowing to pay for increased trade deficits can only postpone or phase out the reduction in consumption and will <u>not</u> enable the countries to avoid the reduction in consumption implied by the deterioration in the terms of trade. The net effect is the same even if the gap is balanced by foreign investment, as the invested capital has to be serviced. And, in the final analysis, investments are made only if the anticipated returns are higher than returns obtainable by merely lending the funds.

The terms under which the capital flows are generated will determine the extent to which the consumption will be reduced. If the capital inflow is a 100 percent grant, the reduction in consumption will be zero, but if the inflow is on fully commercial terms, the consumption will be reduced by the maximum extent. In real life, there is a spectrum of possibilities and the net effect of the capital inflows is not predictable.

Domestic Adjustments

Rational domestic adjustments by reduction in consumption or investment must take into account relative price of goods prevailing in the international market in 1974 and the likely price of such goods in the foreseeable future. Relative price may, in several areas, make production or consumption of substitutes economically more meaningful. In short, the increased price of energy and raw materials will change the comparative advantage of the other developing countries vis-a-vis the other countries in many areas, making many activities economically efficient now. Rational domestic adjustment should move toward harnessing new economic opportunities in all sectors of the economy. We will examine the energy, minerals, agriculture and manufacturing sectors.

Energy Sector

The most promising of the domestic adjustment options appears to be the restructuring of the pattern of fuel production and usage in the other developing countries. Because of the low price of petroleum during the sixties, the average cost of energy was low, the rate of growth of energy consumption was quite high, and the major portion of the increased energy consumption was met by oil. As shown in Table 5, the growth rate of energy consumption was about 5.5 percent in the developed countries, while the growth rate of oil and natural gas was 7.9 percent in the period from

1961 to 1970. In comparison, in the other developing countries the growth rate of energy was 7.1 percent, while consumption of oil and natural gas increased at a rate of 8.5 percent.

TABLE 5

Rate of Growth of Consumption of Fuels During 1961-70
(in percentages)

Countries	Solid Fuel	Oil and Natural Gas	Hydel and Nuclear	Total Energy
1. Developed Countries	0.82	7.92	5.34	5.51
2. Developing Countries	3.21	7.92	10.09	6.87
3. Of which non-oil exporting developing countries	3.28	8.51	10.36	7.05

Source: Computed from the Statistics in World Energy Supplies (1968-71), U.N. (Series J.No.16), World Energy Supplies (1961-70), U.N. (Series J.No. 15).

The consumption of fuel in 1961 and 1970 is shown in Table 6. The table shows the extent to which the share of oil and natural gas in total energy consumption increased in all the countries of the world, the high dependence on oil by the developed and developing countries, and the relatively lower dependence on oil and natural gas by the centrally planned economies.

The dependence of the other developed countries on imported fuels increased during the sixties. The degree of a country's independence in the energy sector can be measured by the "cover coefficient" - the ratio of the country's energy production to its total energy consumption. Among the developing countries only a dozen have cover coefficients of over 50

TABLE 6

Consumption of Fuels in 1961 and 1970
(in million tons coal equivalent)

	Year	Consumption			
		Solid Fuel	Oil & Natural Gas	Hydel & Nuclear	Total
Developed Countries	1961	1034 (39.0)	1548 (58.3)	72 (2.7)	2654 (100.0)
	1970	1113 (25.9)	3074 (71.5)	115 (2.6	4302 (100.0)
Developing Countries	1961	85 (27.2)	220 (70.3)	8 (2.5)	313 (100.0)
	1970	113 (19.9)	437 (76.8)	19 (3.3)	569 (100.0)
Of which non-oil exporting developing countries	1961	80 (33.3)	153 (63.8)	7 (2.9)	240 (100.0)
	1970	107 (24.2)	319 (72.0)	17 (3.8)	443 (100.0)
Centrally planned	1961	914 (74.7)	298 (24.4)	11 (0.9)	1223 (100.0)
	1970	1164 (59.6)	764 (39.2)	23 (1.2)	1951 (100.0)

Source: World Energy Supplies (1968-71) U.N. (Series J.No.16)
World Energy Supplies (1961-70) U.N. (Series J.No.15)

Note: (Figures in brackets represent percentage to total).

percent, and of these all except India have significant production of oil and natural gas. The higher income developing countries have relatively higher cover coefficients, signifying less dependence on oil imports than that of the middle level and lower income developing countries (other than India). (See Table 7)

TABLE 7

Cover Coefficient of Developing Countries in 1970
(In Percentage)

Country	Cover coefficient
1. Non-oil exporting developing countries (all), of (1)	66.6
2. Higher income countries	72.3
3. Middle level income countries	34.2
4. Lower level income of 4	74.9
(a) India	38.9
(b) Others	45.2

Source: Computed from World Energy Supplies (1968-71) United Nations (Series J.No.16) Table 2

Countries with low cover coefficients also seem to have poor indigenous fuel resources; those with relatively higher cover coefficients are better endowed with fuel resources, mostly in the form of oil and gas. In the face of increased oil prices, the countries with higher cover coefficients tend to increase the level of production of fuels within the country. In many cases this may take the form of increased production of coal or electricity (hydel or nuclear), while in some countries, efforts may be made toward accelerating the pace of development of oil or natural gas resources which could not be exploited at pre-1974 prices. Countries with low cover coefficients will concentrate their immediate attention on the energy saving options available to them. In all countries there will be shortrun adjustments to economize fuel, especially oil.

The possibility of reducing the level of energy use in the other developing countries is generally considered to be low in view of the very wide differences that exist in per capita energy consumption in developed and developing countries. In 1970, the annual per capita energy consumption in ton coal equivalent was:

12.5	in North America
3.8	in Japan
4.3	in Western Europe
6.7	in the developed countries as a whole
4.6	in centrally planned economies
0.4	in the developing countries
2.1	in the world as a whole

The average per capita energy consumption in the developing countries was about one-thirtieth that of North America, or one-sixtieth that of the developed countries. In the developed countries, a significant share of energy consumption is used to increase the comfort and leisure of the population - energy used for space heating, for personal transport, for kitchen gadgets, etc. - while such elitist consumption is insignificant in developing countries.

In view of this, the possibility of reducing the level of consumption of energy without a drastic reduction in the level of productive activities appears to be less in the developing countries. This conclusion, however, has to be weighed against the fact that the elasticity of energy consumption to economic growth is substantially higher in the developing countries than in the developed countries.

TABLE 8

Elasticity of Energy Consumption to GDP
1950-1970

		Year	Average rate of growth of GDP	Average rate of growth of energy consumption	Elasticity coefficient
1.	Developed countries	1950-60	4.0	4.1	1.0
		1960-70	4.9	5.3	1.1
2.	Developing countries	1950-60	4.6	7.0	1.5
		1960-70	5.1	7.0	1.4
3.	Centrally planned economies	1950-60	9.4	7.2	0.8
		1960-70	6.7	5.2	0.8

Sources: Statistical Yearbook 1972, United Nations
Statistical Yearbook 1967, United Nations

As seen in Table 8, the elasticity of energy consumption to GDP in the developed countries was 1.04 and 1.1 respectively in the decade of the fifties and sixties; in the developing countries the elasticity was 1.5 and 1.4. This difference is explained partly by the lower efficiency of fuel utilization in the developing countries and partly by the once-over shift from manual and animal power to machines in the early stages of development. It is therefore possible in the lower income developing countries to postpone the pace of mechanization and thus save on energy consumption without affecting the level of production. In all developing countries, if the relative inefficiency of fuel utilization is remedied, it is possible to reduce energy consumption without affecting production plans. But rational reductions in energy inputs which will safeguard the output levels call for a detailed survey of the equipment used in different industries, identifying wasteful practices and equipments, installing improved equipment, and training staffs in more efficient techniques and procedures. Experts widely agree that the possible extent of such reductions by improvement in efficiency would not exceed 10 percent of energy consumption, and that it would take 2 to 5 years to achieve such reductions.

Most of the other developing countries can also substitute indigenous energy for imported energy; such inter-fuel substitution is possible in both the energy producing and energy consuming sectors. In the energy production sector, secondary fuel - mainly electricity - can be produced from any of the primary fossil fuels such as coal or oil or by exploiting hydro-electric potential or nuclear energy. Generally, oil-based electricity generation entails low investment costs but high operational costs, while other modes of power generation require more capital and also involve a longer gestation period. In the case of hydel generation, the scale of production is determined by several location-specific factors. Generally, larger power schemes tend to be more economic than smaller ones, but large hydel schemes call for high technical skills for their design and construction and involve long gestation lags. In spite of the realization that hydel schemes have capital costs comparable to thermal power plants (based on coal) and relatively much lower operational costs in several countries (like India), there is increasing recourse to thermal power generation. The pace of hydel project construction is simply not able to match the rate of increase in demand for electricity.[7]

Nuclear power generation is much more capital intensive than oil-fired power generation and has significant scale economies. In early 1973, the economics of nuclear and oil-fired plants were such that nuclear plants in 800 MW to 1300 MW were definitely more economic for base-load operations.[8/] A preliminary assessment of the impact of high oil prices on the competitive position of nuclear plants vis-a-vis oil-fired plants indicates that even 100 MW plants are competitive if the oil price exceeds $ 9.30 per barrel** while 1000 MW plants can be competitive even if oil prices go down to $ 3.99. (An interest rate of 12 percent was assumed in these calculations.) In all size ranges the capital costs on nuclear plants are about double the capital costs of oil-based power stations of the same size. Given the uncertainties of future oil prices, it would be advantageous to set up large nuclear power plants. But such plants can be set up only in a few developing countries which have power systems with a total generation capacity of about ten times that size. Nuclear power plant design, construction and operation call for high technical skills, which are available only in a limited number of the developing countries. For most of these countries, the foreign exchange requirements of nuclear power generation are bound to be much higher than such requirements for oil-based power generation. Among the developing countries, those in the high income category are most likely to take advantage of the possibilities of substituting nuclear power generation for other fuels, while among the countries in the lower income group, only a few like India with a diversified technological base can opt for this.

There are also possibilities of producing energy from new energy sources like solar energy, geo-thermal energy, wind power and tidal power, and chemical sources. While all these are still in the developmental stage, and investments in them involve large elements of risk, their present development status indicates they are likely to be more capital intensive than the conventional modes of producing energy.

In short, all the options in the energy production sector which will reduce oil requirements are relatively more capital intensive, call for more sophisticated technology and involve longer gestations. While almost all the developing countries will make movements in this direction, only a few in the higher income groups and some technologically more advanced among countries in the other group may be able to take up these options on a significant scale, and even in these countries, the benefits will flow only after a long time.

In the energy consumption sectors, the possibility of economically meaningful inter-fuel substitution must be examined with reference to the purposes for which fuel is used, and the thermal efficiency achievable in each case. In a modern economy, energy is used for:

a) heat-raising
b) lighting
c) providing motive power
d) electrolysis

Electrolysis is solely dependent on the use of electricity, but the other functions of energy inputs can be achieved by using any of the fuels. In certain cases, the use of coal or refined forms of coal, like coal-gas or coke, can be directly substituted for oil while in certain other cases, oil must be replaced by electricity which, in turn, is produced either from coal or by tapping hydel potential or using nuclear fissile material.

The energy using sectors can be broadly classified as household, transport, agricultural and industrial. In the household sector, energy is required primarily for heat raising and lighting; in the transport sector, for providing motive power; in agriculture, for heat raising and providing motive power; and in industry it is required for all purposes. The efficiency of fuel use varies with the sectors, and within each sector the technology available for ready adoption and the price of indigenous fuel may vary with the location. All these factors will have a bearing on the choice of a sector and a purpose in which oil can be replaced by indigenous fuel with the best advantage.

The studies made of India indicate that except for the limited possibility of substituting oil by direct use of coal (as in the case of heat-raising in industries and power generation), the other possibilities of replacing oil by indigenous fuels require the prior transformation of coal to electricity or to coal gas. Such transformations are very capital intensive. Road transport using oil can be replaced by railway transport using electric traction, and considerable savings in oil can be effected. But such a change in the mode of transport becomes economically justifiable only in locations which have high transport density - and the replacement has to be undertaken in a phased manner. It was also found that these options cannot be considered separately as there are inter-relationships among them. The optimal set of interfuel substitutions for each country may have to be determined by simultaneously examining all the options.

An example of such a country study undertaken by a developing country is the Report of the Fuel Policy Committee of India.[9/] This study has sought to determine the optimal pattern of fuel utilization in the light of the current relative prices of fuels. The Committee found that although most of the substitution options in favor of coal were economically desirable, institutional and resource constraints will limit the pace of substitution in the near future. Since the sectors producing energy in primary or secondary forms involve gestation periods of from 5 to 10 years or even more, difficulties in producing adequate substitute fuel, viz., coal, will limit the possibilities of reducing the level of utilization of oil in India in the short run (up to 1978-79). Therefore, if historical growth trends of the economy continue unchanged and only the economically meaningful and physically feasible interfuel substitutions are effected, a substantial shift in favor of indigenous fuels is possible only in the eighties. The likely demand for different fuels, projected on the basis of historical growth trends, and the modified levels of fuel consumption if suitable measures towards optimization of fuel usage are adopted are displayed in Table 9 as given in the report. These provide an illustration of the possibilities for inter-fuel substitution in developing countries. Though the possible reduction in the consumption of oil at the end of the first five years is estimated to be 11.6 percent, it will be 26.5 percent by the end of 1990. The extra investment requirement during the first five years is likely to be about \$ 500 million, and probably a further \$ 300 will be necessary every year thereafter for coal production and power generation facilities. While these amounts are not very large compared to the amounts required to finance the imports of oil which would otherwise become necessary, the long gestation lags for restructuring the pattern of energy utilization will result in a long period during which resources must be found both for the normal level of oil imports and for investment needs.

TABLE 9

Forecast of Fuel Consumption in India
1978-79, 83-84 and 90-91
(million tons coal equivalent)

	1978-79		1983-84		1990-91	
	(a)	(b)	(a)	(b)	(a)	(b)
Coal	135	146	201	218	339	365
Oil	69	61	95	78	155	114
Electricity	70	72	123	128	246	250
Total Energy	274	279[(c)]	419	424[(c)]	740	729

Note: (a) Cols. (a) indicate the likely consumption estimated on the basis of historical trends and capable of sustaining a rate of growth of 5.5% of GNP from 1974-75 up to 1978-79 and 6% thereafter.

(b) Cols. (b) indicate the consumption which can sustain the same levels of economic growth as assumed for Col. (a) computation but with adoption of suitable measures for restraining consumption of oil and for substituting indigenous fuels.

(c) The increase in the total energy consumption on account of the adoption of rationalization of fuel utilization in 1978-79 and 1983-84 is due to the low efficiency of coal using equipment (including power generation plants) compared to the efficiency of oil using equipment.

If we generalize from the studies on the Indian situation, the problem of effecting optimal inter-fuel substitution in many developing countries (with fuel resources other than oil) is likely to be one of finding the necessary investment funds to undertake import substitution projects in the energy sector. The benefits of measures in this direction are likely to be somewhat small in the initial years, and significant only after 5-10 years.

Mineral Sector

The comparative advantage of the developing countries in the mineral sector has changed because of the increases in prices of minerals and metals; it is likely that in most of these countries considerable efforts for modifications in the rate and pattern of production in the minerals sector are being made. In 1970, in the developing countries, the value of minerals produced (net of value of imported minerals) was about $ 7 billion; the degree of processing - defined as the ratio of the value of processed minerals to the hypothetical value if all minerals produced were processed - was only 29 percent. The increased price of raw materials in 1973 and 1974 resulted in a net benefit of about $ 1.5 billion [10] to the developing countries. But the gain accrued mostly to the major exporters of minerals and metals like Chile, Zambia, Liberia, Bolivia, Sierra Leone and Zaire. The majority of the developing countries were net losers because of the increase in the price of minerals and metals. In the face of increased prices, there will be attempts to open new mines or to begin working abandoned mines from which minerals can be produced economically at the current price levels either for exports or for domestic consumption in place of imports.

The possibility of increasing the exports of minerals will be limited by the demand for such minerals in the developed countries. There has been very close co-relation between the rate of growth of GNP of developed countries and the mineral exports from the developing countries. As already discussed, the rate of growth of GNP in the developed countries is likely to be substantially lower in the period up to 1980, as compared to the decade preceding 1973. If the developed countries grow at less than the past rates, the volume of mineral exports to them will be less than would have been planned for in the earlier perspective. Further, the growing awareness of the environmental aspects of mineral utilization and the consequential trend toward designs of equipment and consumer articles which involve progressively less use of metals and lend themselves for easy recycling add another dimension to the market uncertainties. As there is a large gestation lag in mineral development - varying between 5 to 10 years depending on the nature of the mine and the knowledge about mineral deposits already available - a major portion of the future requirements of minerals for the developed countries must have been planned for already. The extent of advance planning is quite significant in mineral and metal production where the industry is dominated by a handful of multi-national corporations with high vertical integration of mineral and metal production. Any reduction in the demand as opposed to earlier expectations is likely to affect the prospective market for mineral producers who do not have links with the multi-national corporations.

While export possibilities for crude minerals are somewhat limited by these considerations, there are some limited possibilities for producing refined minerals or metals for the domestic market, as the developing countries as a whole import refined minerals and metals. Import substitution in the mineral sector in the developing countries will mean beginning further processing of minerals within their countries. In many developing countries this may call for exploitation of low grade ores. The technology required for beneficiating these ores is very sophisticated and will involve heavy investments. Further, such technologies are relatively more energy intensive, a factor which works counter to the moves to reduce the overall level of energy consumption. Even when the capital and technology required for such projects are secured, the uncertainties regarding the future prices in the metal market and the size of the domestic demand relative to the technologically viable size of the plants will make it difficult for the developing countries to proceed rapidly in this direction. Broadly, the domestic adjustments necessary to increase the production of minerals and metals for export purposes is likely to be a feasible proposition only in the few developing countries which have adequate human and technological resources and which can negotiate reasonable agreements with the metal importing countries or the multi-national corporations which are in control of the market. The increase in production of processed minerals or metals for domestic consumption will not generally be possible for the developing countries unless the facilities for these are set up as joint ventures of groups of developing countries. These measures, by their very nature, are likely to be difficult to implement.

It is also possible for the other developing countries to reduce their needs for minerals and metals, although their levels of consumption are already low. The design of household furniture, cooking utensils, and construction techniques provides areas where there can be some savings in the consumption of metals. In quantitative terms such savings will be more significant in the countries which have a comparatively higher level of consumption; thus, the higher income group countries may be able to effect such adjustments in consumption with greater ease than the other developing countries.

Agriculture Sector

The prices of food and beverages have also increased steeply during 1973-74. The other developing countries as a whole are net exporters of food and beverages. In 1970, these countries had a surplus of over $ 5.2 billion on such exports. Of this, $ 2.2 billion accrued to the higher income countries; $ 1 billion to the middle income group and $ 2 billion to the lower income countries. (See Table 10) The exports of the other developing countries consist predominantly of tropical products like sugar, coffee, cocoa and tobacco, while their imports are mostly foodgrains, soybean products and milk products which are produced mainly in the temperate zone. Countries which are major exporters of tropical products appear to be in a position to benefit from these price changes in the long-run. Increasing foodgrain production (wheat, rice, etc.) for import substitution can be attempted in spite of several difficulties.

TABLE 10

Exports and Imports of Food and Beverages of Selected Non-oil Exporting Countries in 1970 (in $ millions)

Country	Exports	Imports	Balance
Other developing countries with:			
i. Higher income	4103	1861	2242
ii. Middle level income	2496	1492	1004
iii. Lower income	3510	1531	1979

Source: Yearbook of International Trade Statistics (1970), U.N.

The agro-climatic conditions, the pattern of land ownership and tenure, the lack of adequate research and development in the production of proper seeds, and the lack of adequate quantities of fertilizers and pesticides, all of which have constrained the foodgrain production in the past, will no doubt continue to operate even in the face of changed prices. But it is relevant to note that during the sixties many of these countries significantly increased the production of cereals by adopting the "Green Revolution" techniques of utilizing new miracle seeds. These techniques were backed up by reliable water supplies from irrigation systems and by the use of chemical fertilizers and pesticides. Still, the average yield of cereals in the developing countries is about 40% of the yield in the developed world. Even this yield is now in danger of being reduced as the price of fertilizer and pesticides, produced mostly from oil, have increased so steeply in 1974, and their supply in the international market is so inadequate that many developing countries find it necessary to cut back on fertilizer utilization. Decrease in fertilizer use, if not compensated by other factors, will reduce the output of food commodities. (It is generally estimated that a ton of fertlizer increases production of wheat by 8 to 10 tons.)

It is in this context that the other developing countries with large populations, and surplus rural labor, may find it economically meaningful to adopt less capital intensive technologies in agriculture. It is possible to replace the capital intensive and relatively more energy consuming methods of ploughing, sowing and reaping used in the "modernised" agricultural sector with methods using more labor. In cattle rich countries like India, where animal dung is used as a fuel, the new price situation can give an impetus to the wider adoption of biogas plants which will provide the fuel value, and also preserve the dung slurry to be used as a nutrient in place of chemical fertilizers. The benefits derivable from this seemingly simple suggestion are enormous. "If all the animal dung is used in biogas plants in India, about 10^{11} million m^3 of 500 BTU/Scft gas can be generated per year, which can meet all the energy needs for cooking for the entire population and at the same time provide about 4 million tons of nitrogenous fertilizers, which is about twice the nitrogenous fertilizer presently being produced from oil products (in India)." 11/ The quicker implementation of long-delayed land reforms in many developing countries will provide an incentive to the cultivators to increase the productivity of the land.

In the agricultural sector, there are possibilities for making gainful domestic adjustments which will mitigate the effects of the increased prices of oil and raw materials. In countries which have the natural endowments for the production of export-oriented tropical agricultural commodities, the tendency will be to increase earnings by suitable adjustments in production. In other countries, which are dependent on foodgrains imports, there are possibilities of increasing production with methods which involve relatively low outlays, like land reforms and adoption of labor intensive but more scientific agricultural practices. Since these adjustments can be made with the greatest advantage in countries with surplus rural labor, several of the lower income developing countries are likely to benefit by making these adjustments.

Manufacturing Sector

The "other developing countries" are not importers of manufactures. Over 70 percent of the net trade deficit of the developing countries is accounted for by the deficits in the trade in manufactures. Of these the higher and lower income other developing countries are more dependent on manufactured imports than the middle level income countries (see Table 11).

TABLE 11

Dependence of Developing Countries on Trade in Manufactures (in $ million)

Countries	Balance on trade of manufactured	Total Imports	Total Exports	Col. 1 as % of Col. 2	Col. 1 as % of Col. 3
	1	2	3	4	5
Developing countries with:					
1) Higher income	-5523	15141	12005	36.5	46.0
2) Middle level income	-2946	9009	8363	32.7	35.3
3) Lower income	-3951	9332	7963	42.2	49.7

Source: Col 1, 2 and 3, as given in Annex Table II
Col 4, 5 computed.

The lower income countries are the most dependent on the importing of manufactured goods.

The prices of manufactures had not increased as much as the prices of primary commodities (see Tables 1 & 2) up to January 1974. There are indications that in the second and third quarter of 1974 prices of manufactured goods have increased rapidly. The price increase in energy, which has pushed up the average cost of energy to the industrial sector, should normally tend to make labor intensive technology relatively more economical in countries with a labor surplus. But in such countries, even at the 1970 relative price of labor and capital or labor and energy, it was economically more desirable to choose technologies in the manufacturing sector which were more consistent with the resource endowments of the countries. However, the factors which hindered the adoption of more "appropriate technologies" in the industries in the developing countries will continue to operate even in the new price situation. The dependence of the developing countries on the developed countries for technology (which moves towards labor-saving techniques) and capital and partially for markets is likely to continue unchanged. The developed countries will evolve more capital intensive and more sophisticated technologies, which could save energy, and raw materials which may not necessarily involve a higher labor input. The capability of the developing countries by themselves to evolve technologies with more appropriate capital-labor or energy-labor ratios being limited, as judged from past experience, the new price relatives do not hold any promise of accelerating the rate of industrialization in the other developing countries.

The developing countries can, however, make attempts toward taking up peripheral industrial activities which can reduce the unit value of their imported manufactures or increase the unit value of their exports of manufactures. On the one side, efforts can be made to increase the value of exports by taking up more "intensive" processing of raw material before it is exported. As already pointed out, the exporters of crude minerals can attempt to set up beneficiation and processing plants. On the other side, imports of fully finished manufactures may be replaced by semi-finished manufactures, like bulk (unpacked) chemicals, machinery in semi-knocked down condition, etc. As the "final touches" stage of manufacture is relatively more labor intensive, these activities can move toward the developing countries where the labor-energy price relative are comparitively low. But the effect of these adjustments is not likely to be substantial.

Impact on Employment

It is tempting to take the simplistic view that, because of the increase in price of energy, more labor intensive technologies should be adopted in the developing countries and employment opportunities should increase as a consequence. The facts are likely to be a little more complicated. If the increased resource gap due to the changed prices of energy and raw material is filled by capital flows and there are no domestic adjustments, there will be no change in the employment situation in the short run. If, however, such capital inflows are inadequate and domestic adjustments are required, the immediate impact on employment will depend on the capital-labor ratio of the new investment compared to the ratio in the sectors from which the investment funds are diverted to new opportunities. Except in the limited opportunities available for domestic adjustments in the rural sector, like adopting labor intensive techniques for import substitution of foodgrains or utilization of animal waste to supply the fuel and nutrient needs of the rural sector, the options are relatively more capital intensive and the net employment opportunities will be less than what they would have been in the absence of price increases.

Summing Up

Given these domestic adjustment possibilities, what choices are likely to be made? Determination of this is difficult. The examination so far points out that any domestic adjustment made would mean reduction in consumption at some point in time because changes in investment patterns would involve opting in favor of higher capital-output and capital labor ratios. These could have serious income distribution effects. Further, the different options involve varying impact on different sections of the community, e.g., drawing away funds committed to the production of synthetic fibers to make investment in nuclear energy production would affect the consumption of certain classes while applying the same funds to biogas plants construction would affect certain other classes. It is, therefore, not merely the resource endowments and technological options that will decide the choices; the ultimate choice of the set of domestic adjustments will depend on the power structure and the political system of each country and will involve value judgments and political choices.

It is, therefore, difficult to classify the countries on the basis of income groups or resource endowments for forecasting the likely adjustments in different countries. In terms of the analysis made in this paper, the broad conclusions that emerge are that the lower income group countries among the other developing countries will find it most difficult to effect even the feasible domestic adjustments, unless they are provided the time to adjust by an increased flow of external assistance. Among the lower income countries, those which have diversified technical skills may be able to adjust their production and trade plans in the energy and manufacturing sectors, while those with large rural populations may effect some appropriate adjustments in the agricultural sector. Some of the countries with mineral resources may be able to take advantage of the new price situation and step up their activities in the mineral sector, provided they are able to obtain technology and marketing assistance either by collective action or by negotiation with developed countries or the multinational corporations which control the mineral industries. The higher income countries among the other developing countries may be the ones who can, through their own efforts, effect domestic adjustments in the mineral, manufacturing and energy sectors. But these are only broad indications of what adjustments are possible. The adjustments that will actually take place depend on factors somewhat beyond the realm of economics. Within each country, there will be long debates before the ultimate choices are made; economists will spend years building their elegant models and scientists will toy with grandiose schemes for the development of new energy resources. But in the meantime the reduction in consumption must be apportioned by politicians among the different classes of people within the limited time scale available if there is no flow of compensating funds in the near future. Many nations will not have the requisite time to pause to plan the optimal choices in domestic adjustments, and some inefficiencies may arise. It is also possible that the violent impact of the changed relative prices will force many countries to choose the same options that they would have chosen even before the increase of commodity prices. In any case, a reduction in the level of welfare is unavoidable for these countries; the basic thrust of international efforts and domestic adjustments should be to ensure that these reductions are no greater than they need to be.

Footnotes

1. Robert McNamara, Address at the Fund Bank, 1974 Annual Meeting, Washington, D.C., September 1974.

2. United Nations: Note by Secretary General: Hypothetical Impact of Commodity Movements on World Trade - DOC/A/9544/Add.2. April 1974.

3. UNCTAD: Note by Secretary General: Problems of Raw Material and Development - DOC/UNCTAD/OSG/52 April 1974.

4. International Monetary Fund: Annual Report 1974, Washington, D.C. 1974.

5. Commission of the European Communities: Towards a New Energy Policy Strategy for the European Community, Brussels, May 1974.

6. Robert McNamara: Address at the Fund Bank Annual Meeting, 1974.

7. In India, the load structure of electricity demand and the characteristics of the hydel projects suggest that ideal power generation system should consist of an appropriate mix of hydel and thermal stations, in which the thermal power stations serve the base-loads, while the hydel stations serve the peak-loads. Even on such system considerations, India should have exploited more of the hydel potential than has been done so far.

8. Lane, J.A.: Kryman R, Raisia, Roberts, "The Role of Nuclear Power in Future Energy Supply of the World" - International Atomic Energy Agency document, Geneva, October 1974.

9. Government of India: Report of the Fuel Policy Committee, August 1974.

10. Computed from the export data in Yearbook of International Trade Statistics 1970-71 and the mineral price index as in January 1974.

11. Ramachandran, A and Bhatnagar: "Energy Research and Development in Asia and Foreign Eastern Countries" - a paper presented at the Third International Conference on Heat and Mass Transfer, Tokyo, September 1974.

ANNEX TABLE 1

Statement Showing Categorization of Developing Countries

	Population (In Million) 1970	Per capital Income (in U.S. Dollars) 1970
DEVELOPING COUNTRIES		
I. Oil Exporting Countries		
1. Algeria	14.33	259*
2. Colombia	21.12	366
3. Egypt	33.33	200
4. Gabon	0.50	468
5. Indonesia	117.89	93
6. Iran	28.66	341
7. Iraq	9.44	278*
8. Kuwait	0.76	3148
9. Libyan Arab Republic	1.94	1450
10. Nigeria	55.07	83*
11. Saudi Arabia	7.97@	344**
12. Syrian Arab Republic	6.25	258
13. Tunisia	5.14	248
14. Venezuela	10.40	854
II. Non-Oil Exporting Countries		
i) Higher level income (Per capita income above $340)		
1. Argentina	23.21	1000
2. Brazil	93.39	368

ANNEX TABLE 1 (Continued)

		Population (In Million) 1970	Per capital Income (in U.S. Dollars) 1970
3.	Chile	8.86	614
4.	Hong Kong	3.96	444***
5.	Lebanon	2.79	521
6.	Mexico	49.09	653
7.	Singapore	2.07	918
8.	Turkey	35.23	352
9.	Uruguay	2.89	787
ii)	Medium level income (Per capita income between $200 and $340)		
1.	Bolivia	4.93	202
2.	Dominican Republic	4.06	334
3.	Ecuador	6.09	250
4.	El Salvador	3.53	274
5.	Ghana	8.64	238
6.	Guatemala	5.28	337
7.	Honduras	2.58	256
8.	Ivory Coast	4.31	321
9.	Jordan	2.31	273
10.	Korea, Republic of	31.02	245
11.	Malaysia	10.40	329
12.	Morocco	15.52	212

ANNEX TABLE 1 (Continued)

		Population (In Million) 1970	Per capital Income (In U.S. Dollars) 1970
13.	Paraguay	2.39	230
14.	Peru	13.59	293
15.	Philippines	36.85	228
16.	Senegal	3.93	201
17.	Vietnam, Republic of	18.83	232
18.	Zambia	4.18	335
iii.	Lower level income (per capita income below $200)		
1.	Afghanistan	17.09	83
2.	Angola	5.58	154**
3.	Burma	27.58	68*
4.	Cameroon	5.84	166
5.	Cuba	8.47	-
6.	Ethiopia	24.63	71
7.	India	539.86	88*
8.	Kenya	11.23	131
9.	Khemar, Republic	6.75	117***
10.	Madagascar	6.75	126
11.	Mali	5.05	50*
12.	Mozambique	7.86	145***
13.	Pakistan	114.18	116*
14.	Sri Lanka	12.51	161

ANNEX TABLE 1 (Continued)

		Population (In million) 1970	Per capital Income (In U.S. Dollars) 1970
15.	Sudan	15.70	109
16.	Thailand	34.38	169
17.	Uganda, Republic of	9.81	127
18.	United Republic of Tanzania	13.27	94
19.	Upper Volta	5.38	57**
20.	Yemen, Democratic	7.21+	96
21.	Zaire	21.57	87

* Data relates to the year 1969

** Data relates to the year 1968

*** Data relates to the year 1963

\+ Population of Yemen and Yemen Democratic.

Source: Yearbook of National Accounts Statistics - 1972, Vol. III International Tables, United Nations
Monthly Bulletin of Statistics - September 1973, Vol. XXVII - No. 9, United Nations

@ The Statesman's Yearbook - 1973/74

Note: Counties in higher and medium level income groups include all non-oil exporting countries with a population of over 2 million and countries in lower level income groups include all non-oil exporting countries with a population of over 5 million.

ANNEX TABLE II

Value of Exports and Imports of Selected Developing Countries: 1970 in 1970 Prices
(In Million U.S. Dollars)

	Oil	Other Raw Materials	Manufactures and Others	Total
EXPORTS				
Developing Countries whose per capita income:				
1. Above $340	420.5	6162.3	5422.6	12005.4
2. Between $200 & $340	198.9	5039.8	3125.0	8363.7
3. Below $200	245.3	5141.4	2576.3	7963.0
Total	864.7	16343.5	11123.9	28332.1
IMPORTS				
Developing Countries whose per capita income:				
1. Above $340	988.2	3206.2	10946.1	15140.5
2. Between $200 & $340	658.4	2279.3	6071.4	9009.1
3. Below $200	797.1	1990.2	6565.1	9352.4
Total	2443.7	7475.7	23582.6	33502.0
BALANCE				
Developing Countries whose per capita income:				
1. Above $340	-567.7	2956.1	-5523.5	-3135.1
2. Between $200 & $340	-459.5	2760.5	-2946.4	- 645.4
3. Below $200	-551.8	3151.2	-3988.8	-1389.4
Total	-1579.0	8867.8	-12458.7	-5169.9

Source: Yearbook of International Trade Statistics (1970-71), United Nations

ANNEX TABLE III

Value of Exports and Imports of Selected Developing Countries: 1970 in 1974 Prices
(In Million U.S. Dollars)

	Oil	Other Raw Materials	Manufactures and others	Total
EXPORTS				
Developing Countries whose per capita income:				
1. Above $340	2434.0	11908.4	7682.7	22025.1
2. Between $200 & $340	1151.3	9739.2	4427.5	15318.0
3. Below $200	1419.9	9935.6	3650.1	15005.6
Total	5005.2	31583.2	15760.3	52348.7
IMPORTS				
Developing Countries whose per capita income:				
1. Above $340	5717.3	7022.6	14854.0	27593.9
2. Between $200 & $340	3809.3	4992.4	8239.0	17040.7
3. Below $200	4611.7	4659.2	8908.9	17879.8
Total	14138.3	16674.2	32001.9	62514.4
BALANCE				
Developing Countries whose per capita income:				
1. Above $340	-3283.3	4885.8	-7171.3	-5568.8
2. Between $200 & $340	-2658.0	4746.8	-3811.5	-1722.7
3. Below $200	-3191.8	5276.4	-5258.8	-2874.2
Total	-9133.1	14909.0	-16241.6	-10165.7

Estimated by using 1974 price ratios* and 1970 values**

* Source: U.N. General Assembly: General A/9544/ADD.2

** Source: Yearbook of International Trade Statistics (1970-71), United Nations

IV. ENERGY USE AND AGRICULTURAL PRODUCTION IN DEVELOPING AREAS

Kenneth D. Frederick

While crop growth involves the conversion of solar energy into digestible energy which can be utilized by humans or animals, agriculture involves the use of human, animal, and fuel energy to manipulate this solar energy conversion. Agricultural systems differ markedly in their use of human, animal, and fuel energy, and these differences are closely related to agricultural productivity. This paper examines these differences and the implications of high fossil fuel costs for the agricultural development prospects of the developing areas.

The Predicament of the Developing Nations

In the developing world agricultural uses of fuel energy generally are either virtually non-existent or very inefficient. Agricultural change in these areas often has come largely in response to increasing population densities, and such changes generally have not meant higher productivity and real incomes.

Where land is not a scarce resource, a forest-fallow system is often employed. The land is cleared by fire, cultivated for a year or two, and then left fallow for 20 to 25 years while nature restores the fertility of the land. Despite the absence of plows and animal power, which are generally incompatible with the forest-fallow system, this system requires surprisingly little labor to provide the basic food requirements of a family.

However, the amount of land needed to satisfy these food requirements is large. As population growth places more pressure on the available land, shorter fallow systems or annual cropping become necessary. Under these systems, the labor required to clear and prepare the land rises sharply in comparison to the forest-fallow system, and unless the farmer "keeps a large herd of domestic animals and uses much labor to collect their manure, prepare composts and spread it carefully in the fields, he is likely to obtain much lower crop yields per hectare under short-fallow systems or annual cropping than by cultivating the same land under the system of forest-fallow."1/ The use of animal power can help limit the increased labor needs associated with shorter fallow systems. However, the animals either have to be fed some of the crop or provided land for pasture.2/ When output is measured in terms of the food available for human consumption and the land input includes the land required to feed the animals, land yields probably decline and the increase in labor yields may be substantially reduced with the introduction of animal power.

After food, the greatest demands on the budget or working hours of many of the world's poor come in providing the energy for cooking and domestic heating. The problems of improving agricultural productivity and satisfying these domestic energy demands are interrelated in the developing countries, and both are aggrevated by population pressures. Wood is the primary source of fuel energy in many areas of the world. While land clearing by fire is a major consumer of energy, the plant energy consumed in areas with sufficient land to employ a forest-fallow system has little or no commercial value and is restored within several decades by simply leaving the land idle. In such areas relatively little effort is required for both farming and providing the wood for heating and cooking. However, few areas of the world still have sufficient land relative to population to support such a farming system. As population pressures force the elimination of forest-fallow agriculture, satisfaction of a developing society's basic cooking and heating energy requirements tends to require more labor and become increasingly competitive with agricultural production for labor and land. Peasants are forced to venture further afield for wood and in some areas wood that a decade or so ago could be collected in an hour now may take a full day's labor. In the Sahel region of Africa 25 to 30 percent of a family's income is commonly spent on firewood.[3/] The uncontrolled use of wood has denuded large areas and contributed to erosion, spread of deserts, and flooding in many densely population areas. In large parts of Asia where the pressures of feeding the human and animal populations have long since eliminated wood as a primary source of fuel, animal dung is the primary source of domestic fuel for rural areas. The burning of manure for fuel, unfortunately, wastes nutrients and organic matter desperately needed to maintain soil fertility.[4/]

Many developing areas are trapped in a vicious circle of population growth, ecological deterioration, and increasing poverty. While the growth of food output exceeded the growth of population for the developing nations as a group from 1952-1972, the opposite was true in 34 developing countries. In about two-thirds of the developing countries the increase in food production was less than the increase in food demand.[5/] Efforts to provide the basic food requirements for growing populations with traditional farming methods and renewable agriculturally-derived energy sources have often resulted in longer working hours and worsening living conditions. And continuation of recent agricultural production trends in the coming decades will greatly amplify the poverty and further undermine the natural agricultural resources of many nations.

Population in the developing market nations is expected to grow at 2.7 percent per annum from 1970 to 2000 and based on the United Nations' medium projection, aggregate food demand for these countries will grow at 3.6 percent per annum.[6/] These growth rates imply increases of 123 percent in population and 189 percent in food demand over the 30 years. Continuation of the rates of agricultural growth achieved in the developing market countries from 1961-1973 would leave their food output falling behind these demand projections by an additional 1 percent per year. By the year 2000 the increase in the production shortfall would amount to 74 percent of 1970 demand levels in the developing market countries. Shortfalls of this magnitude either would have disastrous implications for food consumption or would leave these developing nations with enormous food import requirements.

Even if the developed nations did expand cereal production sufficiently to offset any shortfalls in the developing areas, the obstacles to transferring the grains to the most needy areas would be great. In the absence of an accelerated agricultural development, the income increases required to realize the food demand objectives of the developing countries are not apt to be met. The most vulnerable countries are those dependent on imports of both energy and food products. Such countries would be unable to purchase more than a small fraction of their food deficits on commercial terms, and it is unlikely that food aid would be available for more than a token of the cereal gap.

The only way of ensuring that the food demand objectives of the developing areas are met is to accelerate agricultural development in the developing nations themselves. Agricultural development implies an increase in output per unit of labor or land and usually both, and the introduction of fuel energy inputs has been primarily responsible for the large increases in labor and land yields associated with advanced agricultural systems. In advanced systems, fuel energy becomes very important and takes many forms: mechanization replaces muscular energy inputs in land preparation, planting, irrigation, cultivation, and harvesting; fertilizers help crops convert solar energy into food energy; pesticides reduce crop losses to insects and weeds; and other fuel energy forms are used in the processing and distribution of food.

It is generally agreed that the land and water resources of the developing areas as a whole are capable of supporting large increases in agricultural output. However, replication of low-yielding, low fuel energy consuming agricultural techniques is not possible on a sufficient scale to even keep up with population growth for very long in most developing areas. In many countries, especially those in Asia, agricultural development must be based on land-conserving technologies. In South Asia, for example, over 80 percent of the potential crop land is already under crops. If yields per worker in these areas are not to succumb to diminishing returns, the use of fertilizers, pesticides, and irrigation must expand. All these innovations require much greater use of fuel energy sources than traditional farming methods. In some African and Latin American countries, the agricultural development alternatives are broadened by the existence of large unutilized land resources. For example, land use as a percentage of cropping potential is only 12 in East Africa, 17 in South American and 21 in Central Africa.[7/] Yet, even where land resources are still ample, low-energy agricultural systems can seldom be expanded without encountering declining yields to land or labor. The better lands in terms of accessibility and fertility generally have been cultivated. Either large-scale investments with high-energy requirements will be needed to open the virgin lands or greater on-farm investments will be needed to keep land and labor yields from declining in the face of poorer natural growing conditions. And for some of the largest virgin areas, technological breakthroughs such as improved management of tropical soils and eradication of the tsetse fly are necessary for sustained cropping.

There is considerable uncertainty as to the ability of the developing areas and the costs required to increase agricultural production either by bringing more land under cultivation or by increasing yields on currently cropped lands. These uncertainties have been amplified by the sharp increases in energy prices and the concerns over future energy supplies. The technological alternatives for accelerating farm production and the impact of the energy supply situation on the agricultural development prospects and alternatives in the developing areas are examined below.

Technological Alternatives for Accelerating Agricultural Growth

Among countries with technologically advanced agricultural systems, agricultural development patterns as well as the use of fuel resources vary widely depending largely on the relative endowments of land and labor. Two alternative but not mutually exclusive agricultural development strategies can be distinguished -- those primarily increasing labor yields and those primarily increasing land yields. Agriculture in the United States and Japan typify these alternatives.

As of 1965 agricultural output per farm worker in the United States was nearly 10 times the Japanese output while the output per hectare in Japan was nearly 9 times the U.S. Level. Despite their very different directions, agricultural development in both the U.S. and Japan greatly increased the fuel energy inputs per unit of output in comparison to traditional agricultural societies. The energy inputs, however, take substantially different forms in each country. These differences are illustrated, although oversimplified, by the relative use of tractors and fertilizer. The average farm worker in the United States used nearly 10 times as much tractor horsepower as his Japanese counterpart, while average fertilizer use in Japan was more than 10 times the average U.S. application. In the developing nations tractor and fertilizer use was well below the levels of both the U.S. and Japan. For example, 1965 tractor use in India, Mexico and the Philippines was less than 1 percent of the U.S. average and less than 5 percent of the Japanese average. And 1965 fertilizer use as a percentage of the U.S. average was 16 in India, 11 in

Mexico, and 45 in the Philippines; the average Philippine fertilizer use was only 4 percent of Japan's.[8/]

U.S. agriculture is often cited as a model of efficiency and high productivity. Indeed, the ability of a U.S. farm worker to produce an abundant diet for more than 50 Americans is impressive, especially in contrast to the low productivity of farm workers in the developing areas. However, in comparison to farming in developing areas, U.S. agriculture is a prodigal user of fuel energy. A recent study of U.S. agriculture by the National Academy of Sciences estimates "that farm tractors consumed as fuel about 1.1 percent and farmers purchased as electricity about 0.5 percent of the total national energy consumption in 1972.../and/ the production of fertilizers, chemicals, feeds, machinery, and other inputs purchased by farmers required another 1.9 percent of the total national energy consumption in that year."[9/] While these percentages are not high, the global implications of reproducing U.S. diets and on-farm energy use would be alarming. Pimentel and his colleagues calculated the fuel energy required to provide the estimated 1975 world population of 4 billion people an average U.S. diet using U.S. technology. The annual fuel consumption in gasoline equivalents would average 112 gallons per person for a total global consumption of 488 billion gallons. If petroleum were the only energy source and food production were the only user of petroleum, this rate would consume all known petroleum reserves within just 29 years.[10/]

In the absence of drastic changes in global energy supplies, it is apparent that the developing countries cannot collectively reproduce the farm level energy use of the United States. But even if energy prices returned to their pre-1970 levels and there were no long-run problem of energy supplies, the almost total displacement of manual labor by machines that has occurred in the U.S. would be economically unprofitable and socially disruptive in the developing areas. For the foreseeable future virtually all developing countries must increase agricultural employment as well as output. While this does not rule out the use of tractors for some tasks, over the next several decades farm mechanization in the developing countries should remain well below the U.S. level where mechanization accounts for between 40 and 50 pereent of total (direct and indirect) farm fuel use.[11/] In view of the employment problems confronting all developing nations in the coming decades, it is doubtful that their average tractor use should even reach recent Japanese levels, which in 1965 were about 10 percent of the U.S. average. Farm mechanization should emerge only as overall economic development increases wage rates to levels where tractors become competitive with labor and the displaced labor can be absorbed in other jobs or when selected mechanization can lead to additional agricultural employment by overcoming a particular bottleneck.[12/] Efforts to circumvent or accelerate the overall development process by subsidizing mechanization would be counterproductive even if fuel costs were much lower than they are or likely will be.

Japan's agricultural development is more relevant for most of the development nations than the U.S. experience. Japan has not only succeeded in increasing output and labor yields, it has also increased employment per hectare. In the production or rice, for example, Japan averages about 2.4 times as many workers per hectare as India while maintaining labor yields 1.7 times the Indian average. The primary factors accounting for Japan's success are sophisticated management and use of fertilizers and irrigation. Japan applies from 7.5 to 15 times as much fertilizer to its rice as India, and while all the rice lands are irrigated in Japan, only 40 percent are irrigated in India.[13/]

Unfortunately, the use of chemical fertilizers and ground water irrigation also requires large inputs of fuel energy. For example, fertilizer use in Japan in 1965 averaged 231.7 kg per hectare.[14/] The fuel energy required to produce this quantity of chemical fertilizer would be equivalent to about 37 percent of the average on-farm fuel energy used in U.S. corn production.[15/] Ground water irrigation also requires large amounts of energy. For example, the energy required to apply one foot of water to a hectare of corn in the United States would be equivalent to about 77 percent of the energy embodied in the average fertilizer application in U.S. corn production.[16/] Thus, the energy used per hectare for just the irrigation and the average Japanese fertilizer application would amount to about 65 percent of the total energy used in producing a hectare of corn in the United States.

The technologies which have revolutionized agriculture in the United States and Japan are not unknown or unused in the developing nations. The development of the high-yielding grain varieties which are highly responsive to fertilizer and water underlies what is commonly referred to as the Green Revolution. As the new varieties and recommended input packages were developed for specific regions, adoption of the new technologies was rapid among farmers possessing either lands with the highly favorable growing conditions suitable for the new varieties or the means of improving the growing environment through the application of fertilizer, pesticides, and controlled water flows. Fertilizer consumption in the developing nations as a whole grew at nearly 15 percent per annum from 1967/68 to 1972/73 with much higher rates achieved in areas adopting the new seeds.[17/] Pesticide use has also grown rapidly in the developing areas, and the introduction of power tubewells has been closely associated with the adoption of the high yielding varieties in much of South Asia. In addition, the new technologies have stimulated tractor use in some areas because of the greater importance of and the advantages of mechanization in attaining good soil preparation, seeding at proper depths, and multiple cropping. Large increases in land yields have resulted from the adoption of these technologies; for example, wheat yields from 1960-63 to 1970-73 rose 45 percent in Pakistan and 56 percent in India and rice yields over this same period rose 34 percent in the Philippines, 73 percent in Pakistan, and 14 percent in India.[18/]

In summary the technologies associated with high labor and land yields have required large increases in fuel energy inputs. This has been true regardless of whether the development has occurred in countries with high or low land:labor ratios or in developed or developing countries. Moreover, the fuel energy used in these high yielding agricultural systems has generally come from fossil energy sources which would have to be imported by most developing nations. These characteristics of advanced agricultural technologies are hardly surprising since the technologies were developed under conditions of inexpensive and seemingly abundant energy supplies. However, in view of the dramatic increases in fossil fuel costs and the uncertainties regarding future energy costs and availabilities, the high fuel requirements of these technologies could pose serious obstacles to accelerating agricultural development in the developing areas.

For at least the next decade, agricultural development will depend largely on the rate of adoption of existing energy-intensive technologies; the following section examines the impact of the recent high cost and problems of purchasing fertilizers, pesticides, and fossil fuels on the adoption of some of the technologies most essential to agricultural growth in the developing areas over the next decade. Although the short-term development alternatives are limited largely to fuel intensive technologies, agricultural development does not necessarily imply larger fuel inputs per unit of output. A later section examines the research directions likely both to increase agricultural yields to fuel energy inputs and to lead to economically viable substitutes for the non-renewable energy sources generally used in advanced agricultural systems.

The Supply of Inputs and Agricultural Growth

Adoption of Green Revolution technologies transforms agriculture from a virtually self-sustaining system dependent largely on the current output of natural systems into a system dependent on outside suppliers of inputs produced from non-renewable resources. The increasing dependence on exhaustible resources has important long-run implications for the viability of agricultural systems. Of much greater immediate concern to the rate of adoption of these Green Revolution technologies and the rate of agricultural growth is the resulting dependence of agriculture on inputs which have become much more costly in recent years and over which farmers and often the developing countries themselves have little or no control.

Farming is a risky undertaking and risk aversion is an important objective of farmers, especially poor farmers living on a bare subsistence diet. Traditional farmers are subject to the beneficence of the weather and farming practices are strongly influenced by the need to guarantee a minimum subsistence harvest under a wide variety of conditions. In comparison the Green Revolution technologies are based on the use of seed varieties which respond well to favorable growing conditions. Since adoption of these technologies makes farmers dependent on outside suppliers of seed, fertilizer, pesticides, fuel, and credit, as well as markets for their products, the risk farmers associate with this dependence is an important element in the rate of innovation. The impact of risk and the adequacy of factor markets are complicated by the complementarity between fertilizer, pesticides, irrigation, and good seed. For example, while irrigation is a means of reducing a farmer's dependence on rainfall, the costs of irrigation are seldom warranted unless good seed and fertilizer are available to take full advantage of the water and unless pesticides are available to ensure the crop is not lost to disease or insects. And conversely when water receipts are uncertain because of the lack of irrigation or uncertainties regarding the availability of fuel to operate pumps, the use of fertilizer becomes a more risky and less likely investment.

Both the price and the reliability of access to inputs are important factors in a farmer's willingness to become dependent on purchased inputs. Farmers don't care whether these inputs are imported or are produced domestically as long as the supply is reliable and the price reasonable. However, a nation's willingness and ability to provide its farmers with reliable, low cost inputs is likely to decline if foreign sources are involved. Although reliable supplies of high grade seed are still not common throughout the developing countries, once the seed has been developed by the research stations, the production of good seed is relatively simple and inexpensive, and domestic policies and investments should be able to ensure adequate supplies. Adequate agricultural credit also can be ensured through domestic programs. On the other hand, the availability and cost of the fuel-intensive inputs are commonly determined by foreign suppliers and a country's foreign exchange reserves. The sharp increases in international energy prices and the global fertilizer and pesticide shortages of recent years illustrate the risks of being dependent on foreign fertilizer, pesticide, and energy sources.

India's experience over the past decade illustrates both the tendency for the adoption of new farming techniques to increase dependence on imported fertilizer and energy supplies and the risks associated with this increasing dependence. Nitrogen fertilizer consumption in India grew from 540,000 metric tons (mt) in 1965/66 to 1,487,131 mt by 1970/71; the number of diesel irrigation pumps rose from about 300,000 in 1965 to 600,000-750,000 in the early 1970s, and tractors became a significant source of on-farm power in India's wheat growing areas over this period.[19/]

The growth in the use of these inputs was an integral part of India s success in accelerating agricultural growth in the late 1960s through the adoption of Green Revolution technologies, and the dependence on imports of these inputs was integrally related to their recent setbacks in cereal production.

In recent years fertilizer and agricultural fuel use in India have been slowed by both inadequate supplies and changes in relative prices of farm inputs and products. World fertilizer and fuel prices increased sharply after 1972. For example, urea, a common source of nitrogen fertilizer, increased from about $90 per ton in 1972 to $270 in early 1974 in most Asian ports. [20/] Until 1974, however, government fertilizer subsidies spared Indian farmers from the impact of these price increases. The cost of urea to India's farmers during the 1973/74 agricultural year was only 11 percent above the 1969/70 level while the prices received by farmers increased by 38 percent for wheat and 72 percent for rice over this period. [21/] While these relative price movements substantially increased the profitability of fertilizer use, the consumption of both fertilizer and diesel fuel for the irrigation pumps was restricted by insufficient supplies during the 1973/74 agricultural year. These supply problems are evident in the following quote from India's Economic and Political Weekly:

> In supplies of crucial inputs, such as fertiliser and irrigation from tubewells, the current season is worse off than last year's. Chronic shortage or irregular supplies of often-adulterated fertilisers, is common. For instance, UP's fertiliser requirement was estimated to be 5.5 lakh tonnes; the actual quantity made available--despite the special concessions in the pre-polling period--was only 3 lakh tonnes... 22/

India's fertilizer and fuel supply problems were due to a combination of export restrictions by supplying countries and import restrictions prompted by foreign exchange problems in India. Fertilizer exports from the United States were limited when the Cost of Living Council obtained a commitment from domestic producers to divert more sales to domestic markets, and Japan, which is an important fertilizer supplier to India and other Asian nations, administratively curtailed fertilizer exports in order to meet growing domestic needs. And although the OPEC oil embargo initiated in the fall of 1973 was aimed at a small number of relatively wealthy nations, the poorer nations were also affected by the petroleum shortages and high prices.

The combination of increased domestic production and imports eased the fertilizer scarcity within India by 1974/75. Fertilizer use, however, was depressed by sharp increases in its cost to farmers following the elimination of subsidies in 1974. India's fertilizer use for the year ending March 31, 1975 was down 2 percent from the previous year marking the first decline in eight years. Moreover, the declines in fertilizer use were the sharpest for wheat and rice, the two crops associated with the spread of the Green Revolution in India. Changes in the relative prices of inputs and outputs were particularly important for wheat, the marketing of which was largely taken over by the government. The controlled procurement price of wheat combined with the high prices of crucial inputs such as fertilizer, pesticides, and diesel fuel drastically altered the profitability of using the high yielding varieties and their complementary inputs. As a result of these changes the acreage planted to the high yielding wheat varieties apparently fell in 1974 and the use of fertilizer on wheat fell sharply for the 1974/75 crop year.[23/]

Despite the sharply higher fertilizer costs, the decline in India's fertilizer use probably would not have occurred if product prices had not been depressed by government policies. At free market prices fertilizer was still highly profitable on the high yielding wheat varieties. For example, 1975 farm level prices in India were about $200 per ton of wheat free of government marketing controls and $300 per ton of urea, implying a cost of about $650 per ton of nutrient. Since 1 ton of fertilizer nutrient increases yields by about 7 to 10 tons, fertilizer yielded a 100 to 200 percent return in spite of a 50 percent increase in the price of fertilizer relative to the price of wheat between 1973 and 1975.[24/]

On the global level scarcities and high prices for fertilizer generally will be accompanied by higher prices for agricultural products which will help offset the impact of higher input prices on farm profitability. Since the marginal returns from additional fertilizer use tend to be much higher in the developing than in the developed agricultural nations at current use levels,[25/] the sensitivity of existing use levels to a given change in relative prices might be expected to be less in the developing nations. However, as India's experience suggests, the negative impact of higher international prices and supply problems on input use in the developing countries is likely to be strengthened by other factors. First, for political as well as social reasons the developing countries are likely to be less willing to permit domestic food prices to rise unchecked in response to international price changes. And second, farmers in the developing nations are more likely to have the supplies of inputs dwindle due to the adoption of beggar-thy-neighbor policies by exporting countries or import controls by the developing countries themselves. When the external price rise involves a good with the importance of fossil fuels, a country's import capacity may be seriously eroded, making import restrictions and domestic scarcities a likely outcome for imported products.

The underlying causes of the rapid changes in the energy and fertilizer situations between 1972 and 1974 were fundamentally different but interrelated because of the importance of energy in the production of nitrogen fertilizers. The sharp rise in energy prices between 1972 and 1974 was the result of the actions of the producers' cartel. Current energy prices do not reflect limits in productive capacity or natural resource availability, but unless the cartel collapses, energy prices are likely to remain high for many years. While higher prices for petroleum products contributed to the dramatic rise of nitrogen fertilizer prices beween 1972 and 1974, the most important factor in the price rise was the limited productive capacity in relation to increased demand. As of mid 1974, it was widely believed that the nitrogen fertilizer situation would remain very tight until at least 1980.[26] More recent events and analyses, however, have completely altered these expectations. By the last quarter of 1975 fertilizer inventories were at very high levels, prices had dropped to less than one-half their 1974 levels, and revised estimates of world fertilizer demand and investment indicated substantial excess capacity emerging by 1977. Moreover, much of the new investment is planned for the developing countries and some recent projections indicate that by 1980 the net fertilizer imports of the developing nations will drop to one-half their recent levels.[27]

While the global nitrogen fertilizer supply situation eased dramatically during 1975, farmers in many developing countries may continue to have problems attaining ample supplies at attractive prices in the years ahead. This concern stems from several factors. First, efficient nitrogen plants require large amounts of capital, and some of the investments planned during recent periods of high fertilizer prices may never be realized. Second, at the lower price levels, fertilizer consumption in the developed countries can be expected to rise sharply in comparison to the depressed 1975 levels. Third, the technology involved in modern efficient nitrogen fertilizer production is very complex and developing countries have a poor record in bringing plants into operation on time and operating them near capacity. And fourth, a small group of countries organized and committed to exploiting their monopolistic power have considerable control over the feedstocks for nitrogen fertilizer plants.

Farmers have also been confronted in recent years with tight supplies and sharply rising prices for phosphate fertilizers. These changes have been the result of efforts to exploit monopolistic power and not capacity constraints. In particular, in a series of price changes totally unrelated to either production costs or supply shortages, Morocco, the principal exporter, increased phosphate rock prices 4.5 fold between late 1973 and July 1, 1974.

Potash, the third major component of chemical fertilizers, has not been subject to market disruptions similar to those for nitrogen and phosphate, and no disruptions are anticipated. However, global potash supplies are dominated by a single country, Canada, and within the current environment where many suppliers are seeking and exploiting monopolistic power, the concentration of potash supplies also may have a negative impact on agricultural development decisions.

Most developing nations are also dependent on foreign suppliers for pesticides. A United Nations document prepared for the 1974 World Food Conference estimated a worldwide deficit of 20-30 percent for pesticides in 1974-75, a situation they anticipated would almost eliminate supplies to the developing nations.[28/] While this assessment proved to be overly pessimistic, recent price increases and shortages emerged in part because of the dependence of many pesticides on petroleum feedstocks and the diversion of these feedstocks to other uses. Over the longer term, the developing nations can strengthen their pesticide and chemical industries if they have access to the required technology. Nevertheless, agricultural expansion is likely to leave most developing nations increasingly dependent on imports of pesticides or the basic feedstocks for preparing pesticides.

In summary, while the developing nations have great potential for increasing agricultural production, increasing fertilizer, pesticide, and energy consumption is a *sine qua non* for accelerating agricultural growth over the next decade or two. Adoption of these inputs, however, implies a substantial increase in risk for individual farmers as well as for the developing countries because of the dependence on imported inputs. If this dependence is viewed as a further threat to the success of adopting new varieties and farming methods, innovation in the developing areas will be slowed. While astute planning and international cooperation can assure adequate global fertilizer and pesticide supplies, the ability of many of the poorer nations to provide their farmers with reliable supplies of these inputs depends on the cooperation of the oil exporting nations. The OPEC countries as the owners of the least expensive feed stocks for nitrogen fertilizers and huge quantities of capital have an important role to play in ensuring the needed fertilizer investments are made. Yet these investments will do the poorest countries little good unless they have the foreign exchange to purchase the needed inputs. The import capacities of the oil importing, developing countries have been hardest hit by the sharp rise in petroleum prices, and the prospects for agricultural and overall development in the poorest countries are bleak unless most of the impact of the higher oil prices is offset by increased flows of foreign investment or aid.

Long-run Problems and Possible Solutions

The high degree of dependence of advanced food systems on a finite supply of fossil fuel and the drastic increases in fossil fuel costs in recent years suggest both the need for and the economic desirability of developing and adopting alternative agricultural technologies. New technologies are needed in the developed agricultural areas to maintain historical growth rates of yields since the technologies underlying prior growth and current high productivity may be approaching their natural limits. 29/ And while the developing areas still have great potential for agricultural expansion through the adoption of fuel-intensive technologies, this growth may lag if it remains dependent on existing technologies and traditional sources of fuel energy. Technologies requiring less fuel per unit of output and alternative energy sources for farmers are needed to ease the food and energy problems of the developing nations in the coming decades.

While technologies with considerable potential for increasing land and labor yields without larger fuel energy inputs have been under development in the United States and several other countries for many years, such research has received only a token of the support provided other agricultural research. Two of the most promising research areas seek to enhance the photosynthetic efficiencies of plants and the abilities of plants to fix nitrogen in the soil. 30/

The underlying determinant of crop yields is the degree to which plants convert solar energy into plant material. Fuel-intensive technologies increase yields to land by using fertilizers and controlled water receipts to improve the growing environment of crops. Biological development of new seed varieties which respond particularly well to conditions of high fertility and abundant water underlie the most dramatic yield increases obtained to date. Another line of research seeks to improve the efficiency with which plants photosynthesize in a given environment. Plants are inefficient converters of solar energy into food energy. Photosynthesis seldom converts more than 1 percent of the sunlight available during a growing season into plant growth, and in general only about 15 to 20 percent of the theoretical limit to the conversion of solar energy into plant material is achieved. Even the most efficient crops producing record yields now photosynthesize at only about half their potential. However, science may be able to improve upon this record. Major scientific breakthroughs in the understanding of the photosynthetic process achieved during the 1950s have been followed by research designed to increase the photosynthetic efficiency of some plants. The optimism some scientists have for such research was expressed by Wittwer as follows:

> There is no research area where the opportunities are more attractive and the potentials greater for achieving results, reflected in increased crop productivity, than in maximizing the photosynthetic process. 31/

Regarding the types of changes that might improve the efficiency of plant photosynthesis, Wittwer states:

> With all plants there are three possible complementary and parallel routes - select genetic variables with greater photosynthetic efficiencies, modify plant architecture for better light reception, and apply chemicals to suppress glycolate biosynthesis and inhibit photorespiration. The balance between photosynthesis and respiration can be chemically, physically, and genetically altered to maximize productivity. 32/

Only the first two routes would clearly increase the yields to fuel energy inputs. The application of chemicals to improve photosynthetic efficiencies, such as the use of carbon dioxide gas in greenhouses or dry ice in open fields, is expensive and energy intensive. Although carbon dioxide applications have resulted in much higher yields for some crops, much more efficient application techniques must be developed to be commercially viable for crops in open fields. 33/ And even then the innovations requiring chemicals may not increase crop yields per unit of fuel energy.

Increasing biological nitrogen fixation is another promising area for reducing the fuel energy requirements of high productivity agriculture. Nitrogen fixation involves the extraction of nitrogen from the air by bacteria working in a symbiotic relationship with plants. This nitrogen becomes a nutrient for plants reducing the need for fertilizer. Legumes possess this ability to fix nitrogen in the soil, and research is attempting to increase the amount of nitrogen fixation and the yields of some legumes such as soybeans. Moreover, recent findings suggest that the nitrogen fixing ability might be genetically transferred to crops such as the grains that do not now have this characteristic. By reducing the need for chemical fertilizers, success along these lines could dramatically reduce the fuel energy requirements of food production.

As was noted earlier, wood and animal dung are the principal sources of fuel energy for most of the world's people, and domestic cooking and heating are the primary uses of energy in poor rural areas. The predicament of farmers with these limited energy demands and sources is desperate and often deteriorating. Many are having to work harder just to maintain a subsistence income and the chances of failure are increasing. While such farmers do not have machines to ease the burden of farming, the collection of wood for fuel is placing increasingly large demands on the farmers time and energy in many areas. Moreover, the use of wood and animal dung for fuel is undermining the agricultural land resources in many areas. Unless these areas can convert to alternative energy sources or eliminate the ecological costs associated with current fuel use, these environmental and poverty problems will be extended and aggrevated by population growth.

Conversion to alternative energy sources generally has implied rural electrification or the use of some form of fossil fuel to power motors and provide heat. Indeed, agricultural development undoubtedly will continue to be associated with development of these power sources in many areas. However, the high costs both of fossil fuels and of providing electricity to small villages will dampen such an expansion and make these alternatives impractical for many rural areas. Other alternatives are essential for the economic development of many relatively isolated rural areas.

Wind has long been used for pumping water, and the pressing needs for increased irrigation and the global energy problems may expand the role of windmills. While the production of electricity from the power of the wind or sunshine is not currently economical, research and higher prices for conventional energy sources could alter this in the coming decades. Currently, however, biogas plants are a much more promising alternative for satisfying the energy needs of communities too poor or small to warrant the transmission of power from centralized plants. 34/

Although it is not now practical to use wood or agricultural wastes directly to generate steam for power production, biogasification is a practical means of converting biological materials other than wood to more useful forms of fuel. Biogasification involves the breakdown of materials such as plant residues and dung by anaerobic bacteria into compounds which are in turn converted into methane and carbon dioxide. The residue which remains undigested provides an excellent organic fertilizer. While composting can capture the fertilizer potential of organic wastes, the fuel potential of the wastes is lost with composting. Biogasification, on the other hand, results in no significant loss of fertilizer value and also produces a form of energy which is usable for irrigation pumps, farm machines, trucks, or cooking. Biogasification, of course, involves capital costs not encountered in composting, and in areas where plant wastes left on the ground help curtail erosion, there may be ecological costs in collecting them for use in biogas plants. However, under a wide variety of circumstances encountered in developing areas, these costs may be offset by the value of the fuel and fertilizer produced. While biogas plants are already in use in some areas, especially in India, their advantages would appear to warrant a more rapid and widespread construction of such plants.

The energy potential of methane derived from the biogasification of plant residues and animal dung is great, in fact more than enough to enable the agricultural sector to be self-sufficient in energy. Even with the energy intensive agriculture of the United States, the energy potential of biogasification exceeds agricultural energy use. 35/ Moreover, the biogas potential increases with agricultural expansion.

A variety of other changes in farming practices or biological innovations might also enable farmers to increase yields without increasing their use of energy. Biological methods of controlling plant pests and genetic plant selection may replace pesticides in many uses providing environmental as well as energy saving benefits. Experiments suggest that a switch from conventional mechanized cultivation methods to minimum or zero tillage can reduce the fuel requirements of tractor use by 40 to 80 percent.36/ A study of agriculture in the United States by the National Research Council conjectures that "Aerial instead of tractor application of seed, fertilizer, or pesticides might reduce energy consumption." They add, however, that "Knowledge of the energy savings attributable to these alternatives is lacking." 37/ Adoption of improved crop rotation schemes or composting, in areas where the plant materials and dung are not used as inputs for biogas plants, can reduce the need for chemical fertilizers. And mulching with a wide variety of materials readily available in rural areas can reduce irrigation and cultivation requirements. In areas where land is scarce and wood is a major source of fuel, wood plantations may be able to supply the fuel requirements on only one-fifth of the land required with the practices generally used. 38/

Summary

The most productive agricultural systems make extensive use of fuel energy inputs. Mechanization increases labor productivity many fold while other fuel-intensive inputs such as chemical fertilizers, pesticides, and irrigation facilitate the growth and reduce the losses of crops. The least productive agricultural systems make little or no use of fuel energy inputs. All too often the combination of farming, grazing, and the use of wood or animal dung for domestic fuel requirements precipitates erosion, flooding, and declining soil fertility. With population growth placing increasing stress on the available resources, this ecological deterioration and the resulting poverty cycle can only be broken through a more extensive or a more efficient use of fuel energy inputs.

Two alternative but not mutually exclusive agricultural development paths are represented by the experiences of the United States and Japan. The high degree of mechanization in the United States makes U.S. farmers among the most productive in the world, while very high use levels of chemical fertilizers and irrigation in Japan make their land yields among the highest anywhere. In view of the low cost and abundance of labor in the developing areas, Japan's agriculture provides a much more relevant model for the developing areas.

The agricultural technologies characteristic of Japan and the United States are not unknown in the developing nations. Indeed, the tremendous enthusiasm and expectations that led to the widespread acceptance of the existence of a Green Revolution in the developing areas in the late 1960s was prompted by important progress in the adoption of these technologies. The potential for increasing agricultural production in developing areas through the adaptation and adoption of these technologies is great. Recently, however, the spread of these techniques has been threatened by the high cost and uncertain supplies of essential inputs. Adoption of the Green Revolution technologies leaves farmers dependent on outside suppliers of such inputs as high-yielding seed varieties, fertilizers, pesticides, and fuel for irrigation. This dependence together with the high complementarity of these inputs on agricultural production makes their adoption a risky undertaking in areas without well established product and factor markets. The combination of erratic prices and uncertain supplies of fertilizers, pesticides, and fuel has slowed the expansion of Green Revolution techniques in India and contributed to much more pessimistic prognoses regarding the applicability and spread of the Green Revolution technologies in the developing areas generally.

If the food demand projections for the developing nations are to be met over the next several decades, these countries must accelerate their agricultural growth rates. The availabilities of land and water resources and agricultural technologies suggest that, for at least the next decade, agricultural development will depend largely on the rate of adoption of existing energy-intensive technologies. Fossil fuel prices are not likely to return to their levels of the 1960s within the foreseeable future and undoubtedly will have an inflationary impact on the prices of other key agricultural inputs. Nevertheless, the fertilizer and pesticide situations have eased substantially since 1974, and even at 1974 input prices, fuel-intensive technologies are highly profitable for many farmers as long as crop prices are not artificially reduced.

For the longer run, new technologies are needed to reduce the dependence of advanced food systems on fossil fuels. Research to improve the photosynthetic efficiency of plants and the abilities of plants to fix nitrogen in the soil offer hope for improving agricultural production without increasing the fuel energy requirements. However, widespread commercial application of these major fuel-saving innovations appears to be at least a decade or two away. Of more immediate relevance for some of the world s poorest rural communities is the production of fuel and fertilizer through the biogasification of plant residues and animal dung. In areas where farming combined with the use of wood or dung for domestic heating and cooking are resulting in ecological deterioration, biogas plants offer a viable means of increasing the production of food and fuel and curbing the ecological damage of current practices.

Footnotes

1/ Ester Boserup, The Conditions of Agricultural Growth: The Economics of Agrarian Change Under Population Pressure (Chicago: Aldine Publishing Co., 1965), p. 32.

2/ Ibid., p. 35 notes that "In most parts of the world, the area an animal can till with a primitive plough is much smaller than the area needed to feed it on natural grazing."

3/ Eckholm, "Cheaper Than Oil, But More Scarce," The Washington Post (Sunday, July 27, 1975), p. C3.

4/ Ibid., feels that the loss of nutrients when people are forced to resort to burning cow dung is probably "the most pernicious single result of firewood scarcity."

5/ Food and Agriculture Organization, The State of Food and Agriculture 1974 (Rome, 1974), pp. 93-94.

6/ United Nations, Assessment of the World Food Situation: Present and Future, Item 8 of the Provisional Agenda of the World Food Conference November 1974, table 14, p. 79. The UN projections for 1985 and 1990 have been extrapolated to 2000. In view of the low initial consumption levels in the developing areas and the unlikelihood that the population levels will be significantly altered in the absence of a major world-wide disaster, these food demand growth rates might be viewed as minimum objectives.

7/ These data on land use and cropping potential are from the Economic Research Service of the USDA.

8/ The data on land and labor yields and fertilizer and tractor use are from Hayami and Ruttan, Agricultural Development: An International Perspective (John Hopkins University Press, 1971), p. 70 and appendix table A-5.

9/ The National Research Council, Agricultural Production Efficiency (Washington, D.C.: National Academy of Sciences, 1975), p. 119.

10/ David Pimentel and others, "Food Production and the Energy Crisis," Science (Nov. 2, 1973). vol. 182, no. 4111, pp. 447-448.

Footnotes (cont.)

11/ According to estimates in the National Academy of Sciences study mentioned above, the fuel consumed for tractor use accounts for 31 percent of the total fuel energy used either directly or indirectly to produce crops in the United States. Estimates by Pimentel and others of the average energy inputs in U.S. corn production in 1970 attribute 27.5 percent to direct consumption of gasoline and 14.5 percent for the construction and repair of tractors, trucks, and farm machinery. (See Pimental et. al., "Food Production and the Energy Crisis," Science (Nov.2, 1973). pp. 444-445.) Since the corn energy estimates are based on fertilizer applications which are considerably above the average for all U.S. farms, the 40 to 50 percent range appears to be a reasonable approximation of the importance of mechanization in on-farm energy use.

12/ For example, double cropping in some areas may require mechanization for rapid harvesting and land preparation.

13/ Arjun Makhijani in collaboration with Alan Poole, Energy and Agriculture in the Third World, a report to the Energy Policy Project of the Ford Foundation (Cambridge, Massachusetts: Ballinger Publishing Co., 1975). table 1-2, p. 8.

14/ Hayami and Ruttan, appendix table A-5.

15/ The data of Pimentel et. al, indicate that in the U.S. an average of 203 pounds of fertilizer are applied per acre of corn. This amount, which is equivalent to 227.4 kg per hectare, accounted for 36.4 percent of total on-farm energy use associated with U.S. corn production.

16/ Calculated from data in Pimentel et al, pp. 444-445. The energy for ground water pumping depends on the pumping depth.

17/ United Nations, Assessment of the World Food Situation: Present and Future (1974), table 6, p. 40.

18/ Dana G. Dalrymple, "The Green Revolution: Past and Prospects", Economic Research Service of the USDA and Bureau for Program and Policy Coordination of USAID, Draft no. 2 (Washington, D.C., July 1974), p. 21.

19/ Ralph W. Cummings, Jr., Food and the Energy Dilemma, working papers, The Rockefeller Foundation (November 1974), pp. 13-15.

20/ Ralph W. Cummings, Jr., An interview in Ceres, (May-June 1974), p. 37.

Footnotes (cont.)

21/ Economic Research Service, USDA, World Fertilizer Situation 1975, 1976, and 1980, WAS 5-Supplement (October 1974), p. 36.

22/ Economic and Political Weekly, vol. IX, no. 10 (March 9, 1974), p. 387.

23/ The Farmers' Forum, which has attempted to discourage wheat planting as a protest against the government's intervention, claims wheat planting fell 20 percent in both Punjab and Haryana. The Economic and Political Weekly, vol. IX, no. 10 (March 9, 1974), p. 387.

24/ The data on fertilizer and crop prices in India are from John Parker of the Economic Research Service, USDA.

25/ Raymond Ewell has estimated that the marginal grain-fertilizer response ratio is about 10 in developing countries and 5 in the United States. See his statement before the U.S. Foreign Affairs Committee of the House of Representatives, Ninety-third Congress, 2 sess. (September 1974). Reprinted in U.S. Policy and World Food Needs (Washington: USGPO, 1974), p. 70.

26/ See for example, United Nations, The World Food Problem: Proposals for National and International Action, Item 9 of the Provisional Agenda of the World Food Conference, (Rome, November 1974).

27/ Based on personal communication with Richard Reidinger, ERS, USDA, in November 1975.

28/ United Nations, The World Food Problem: Proposals For National and International Action, p. 50.

29/ Agricultural productivity prospects for the United States are examined in National Research Council, National Academy of Sciences, Agricultural Production Efficiency,(Washington, D.C.: 1975).

30/ The following discussion of photosynthetic efficiency and nitrogen fixation draws heavily on two sources: S. H. Wittwer, "Maximum Production Capacity of Food Crops", BioScience, vol. 24, no. 4, (April 1974), pp. 216-224, and Gene Bylinsky, "A Scientific Effort to Boost Food Output", Fortune (June 1975), pp. 99-105, and 172-173.

31/ S. H. Wittwer, "Maximum Production Capacity of Food Crops", BioScience, vol. 24, no. 4, (April 1974), p. 216.

Footnotes (cont.)

32/ Ibid., p. 217.

33/ Gene Bylinsky, "A Scientific Effort To Boost Food Output", Fortune (June 1975), p. 102.

34/ The following discussion of biogasification if based on material in Makhijani, chapter 4, which provides a detailed examination of the conditions under which biogasification would be a competitive source of energy for agriculture.

35/ Makhijani, pp. 95-96.

36/ Food and Agriculture Organization of the United Nations, The State of Food and Agriculture 1974 (Rome: 1974), p. 30.

37/ National Research Council, Agricultural Production Efficiency, p. 127.

38/ Makhijani, p. 69.

V. ENERGY AND THE LESS DEVELOPED COUNTRIES: NEEDS FOR ADDITIONAL RESEARCH

Alan Strout[1/]

In the wake of the recent three-fold increase in world oil prices, two questions have become of major importance to those countries in the early stages of economic development. How will future energy supplies be obtained and paid for? How important is energy to economic development and what are the prospects for reducing the apparent historical dependence of economic growth on energy? This paper deals with these two questions from the viewpoint of gaps in our understanding of the relevant technological and economic relationships.

Special Characteristics of Energy

There are several peculiarities of energy as a productive resource which influence its use during the course of economic growth. First, there is a great deal of relatively low-grade energy widely available (wood, vegetable matter, water power, and wind), and this has led to the rather dispersed and low-density economic activities, largely agricultural, of pre-industrial man. Modern economic development has paralleled the discovery and exploitation of new energy resources (chiefly coal, petroleum, and natural gas) which have permitted much higher concentrations of use, both in time and place. Economic activities have become more concentrated, permitting large economies of scale and organization. The story of urbanization has been the story of increased energy use.

These utilities of form and time and place continue to be important and account for the 10:1 price difference on a heat-equivalent basis between fuel wood and gasoline. One of the options for developing countries, however, may be to more fully exploit the low grade energy already in their possession and to evolve a more geographically dispersed development pattern than has historically been found among the currently more advanced countries.

Second, the use of energy in all but the most primitive society of hunters and gatherers is inextricably linked to physical capital formation. Energy must be channelled, contained, and controlled. Physical structures or machinery are therefore required. To capture solar energy through crops, land must be cleared and often leveled and irrigated. Transportation requires vehicles. Factories require machinery and equipment for the control or containment of energy, and factory buildings permit other energy to be used as heat, light, or air-conditioning for the comfort of the human workers.

In passing through the physical system for control and containment, however, a great deal of energy is lost. Efficiencies of energy use are generally low, and this is to some extent a function of the capital equipment itself. It is likely that as a result of increased energy prices, increased investment in machinery and equipment (and perhaps in the organization of human activity -- also a form of "capital" investment) will lead to higher efficiencies of energy use. We know little about substitution possibilities between energy and capital, however, especially when these may involve complete systems of interrelated energy-using activities.

Finally, most economic activities require energy, and energy thus takes its place along with capital and human labor as one of our universal productive factors. In important ways, however, energy is not different from other natural resources. There is considerable room for substitution among the many forms of energy (although the accompanying capital investment costs may be high). Thus when one form of energy becomes scarce or higher priced there is a good chance that it can be substituted for by some alternative combination of capital and another form of energy. Technological change and chance discovery also play an important role in energy supply as they do in the supply of other materials. New reserves may be discovered at any time and the very definition of "reserve" is constantly affected by technological change.

Preliminary Orders of Magnitude

What do we know about the historical relationship between economic growth and energy? It turns out that most of the differences in energy use observable today among countries are related to current differences in GDP (or GNP) per capita and to the size of the country measured by its population. In general, "commercial" energy use (fossil fuels, hydro, geothermal, and nuclear power) is proportionately higher at higher levels of per capita GDP and is proportionately less, other things being equal, for countries with large populations. Table 1 gives the "standard" energy consumption pattern at various income levels for countries of 10 and 100 million inhabitants. Actual and calculated values for India, Pakistan/Bangladesh, Brazil, Italy, Japan, and the U.S.A. are shown for comparison.[2/]

Following United Nations practice, energy is measured on a more-or-less comparable basis of "coal equivalence" and in metric units. These units have been abbreviated as MTCE (metric tons coal equivalent) or KCE (kilograms coal equivalent).

When fuelwood (as reported by the FAO) is added to commercial energy, the patterns remain similar to those shown in Table 1. At a level of $125 per capital GDP, fuel wood adds about 160 KCE/person to total energy use of the smaller size countries and about 140 KCE/capita to energy consumption in the "standard" country of 100 million inhabitants. These differences, however, disappear at a per capita GDP level of above $1,000.

It appears, furthermore, that the country-size effect noted in Table 1 may largely reflect price differences which distort the conventional measures of Gross Domestic Product. Extrapolating from a careful 10-country study by Gilbert Kravis, Zoltan Kenessey, Alan Heston, and Robert Summers,[3/] crude GDP price deflators can be estimated for each of the 23 countries in our 31-country sample which were not covered by Kravis and his colleagues.[4/] When reported GDP for each country has been adjusted to a common price basis using the crude GDP deflators, both the population effect and the squared-GDP effect became statistically insignificant. This was true whether or not fuel wood was included. The purchasing-power-adjusted income elasticity for commercial fuels was about 1.6 and for the total including fuel wood, about 1.3. Since our conventional thinking about economic growth is in terms of exchange-rate GDP rather than purchasing-power-equivalent measures, however, and since a great deal of more work is needed before we fully understand the reasons for purchasing

TABLE 1

Energy Consumption at Various Levels of Development

I. Hypothetical countries conforming to "standard pattern"[a/]

GDP per Capita, 1970 US$	Commercial Energy Consumption, KEC/person	
	10 million inhabitants	100 million inhabitants
125	132	80
250	473	288
500	1390	848
1000	3350	2040
2000	6590	4020
4000	11900	7240

II. Representative actual countries, about 1970

Country	GDP/capita, US$ (UN)	Population, millions	Commercial Energy Consumption, KEC/person	
			actual[b/]	hypothetical, from standard pattern[c/]
India	99	550.4	180 (192)	137
Pakistan/Bangladesh	176	129.0	93 (99)	172
Brazil	402	95.3	480 (585)	771
Italy	1724	53.7	2590 (2829)	3330
Japan	1900	103.4	3108 (3322)	3990
U.S.A.	4832	204.8	11025(11388)	9890

[a/] Derived from a statistical relationship covering 31 countries. The estimated equation used was:

$$\ln EN/POP = \underset{(4.41)}{-13.78} + \underset{(4.31)}{3.996} \ln GDP/POP - \underset{(2.93)}{.208} (\ln GDP/POP)^2$$

$$\underset{(3.32)}{-1.645} \ln POP + \underset{(3.65)}{.207} (\ln POP)^2$$

where t-ratios are given in parenthesis

$\bar{R}^2$ = .930 and SEE = .404 (which is equal to a mean standard error of approximately ± 50% of the value of the untransformed dependent variables)

EN = Gross energy consumption, kilograms x10^6 of coal equivalent, from U.N. World Energy Supplies, 1969-1972, Statistical Papers Series J. No. 17 (New York, 1974), average for 1969-1971.

POP = 1970 population x 10^6, U.N., Monthly Bulletin of Statistics, November 1971.

GDP/POP = Gross Domestic Product per person, U.S. $, average for 1969-1971 where possible, from U.N., Monthly Bulletin of Statistics, January 1975, Special Table E.

b/ The first number shows energy consumption calculated by the United Nations. Imported electricity or that produced by hydro, nuclear, and geothermal sources is included at the direct calorific heat value of the energy itself. It is this value that has been used for the statistical regression reported above. The estimate in parentheses includes an adjustment to express hydro, nuclear, and geothermal electricity and imported electricity at a common "thermal fuel generating" equivalent of 10494 Btu/kwh. It is this latter figure which is generally referred to in the text as "total commercial energy."

c/ From "Future of Indonesia" computer run 5/6, equation 8, of June 25, 1975.

power differences among countries, we will continue to refer to the standard pattern based on nominal per capita income and shown in Table 1.

The significance of Table 1 lies in the disproportionately large amounts of energy which appear to be needed at higher levels of per capita income. If the growth path of an individual developing country should follow that long-tern pattern described by the cross-country results (and many an event may cause the two patterns to differ in reality), a four-fold increase in per capita GDP, from $125 to $500, would require a ten-fold increase in commercial energy consumption.

Assuming a 1970 average energy price level of about $25/MTCE (roughly $30/MTCE for petroleum products and $20/MT for coal itself), the cost of this commercial energy would increase from 1.6% to 4.2% of GDP for a country of 100 million population and from 2.6% to 6.9% of GDP for a 10-million population country (Table 2). These are relatively modest energy costs, and the difficulty comes when (a) energy prices become sharply higher and (b) the bulk of commercial energy must be imported. At an average cost level of $70/MTCE ($90/MTCE for petroleum products and $50/MT for coal), the preceeding percentages would increase by a factor of 2.8. Assuming that a country is self-sufficient in energy and the higher prices represent increased rents to nationals of the country itself, the cost of the energy to society as a whole is not changed. If fuel is owned by foreign nationals and must be imported, then the higher cost becomes a significant charge against the country's earnings of foreign exchange.

This can best be illustrated by calculating a country's "normal" pattern of export earnings (estimated by Hollis Chenery and Moises Syrquin using a model similar to that incorporated in Table 1). These are shown in Table 3. A country of 10 millions at low per capita income levels tends to export about twice as much, proportionately, of its gross product as does one of 100 millions. This difference increases at higher income levels. If we assume, to take an extreme but not unknown example, that almost all commercial energy is owned by foreigners or must be imported, then the 2.8-fold price increase assumed above would have to be matched by a 22 percent increase in export earnings for a smaller country at low income levels (assuming that non-energy imports were not affected). At a GNP level of $1000 per capita, export earnings for the 10 million population country would have to increase by 50 percent to finance the higher-priced energy imports.

The problem is most acute in absolute terms for the smaller countries because energy use per capita tends to be higher for a given income level (ignoring differences in purchasing power). These countries are generally oriented to foreign trade, however, as judged by relatively high export and import levels. Smaller countries may thus find it relatively easier to expand their export earnings than do larger, more self-contained and less open countries. Relative increases in export earnings for a country of 100 million inhabitants, for example, would reach 77 percent at an average GDP per capita level of $500 in order to compensate for the higher cost of energy imports assumed in this example.

The standard statistical inter-country patterns suggest that energy use increases less rapidly than GDP at income levels above $1000 per

TABLE 2

Costs of Energy Relative to GDP, Various Income Levels, Small and Large Countries, Assuming Common Development Patterns and Energy Prices

Per Capita GDP, 1970 US $	Energy cost of $25 per MTCE[a/]		Energy cost of $70 per MTCE[a/]	
	10 mil. Pop.	100 mil. Pop.	10 mil. Pop.	100 mil. Pop.
125	.026	.016	.074	.045
250	.047	.029	.132	.081
500	.069	.042	.195	.119
1000	.084	.051	.234	.143
2000	.082	.052	.231	.141
4000	.074	.045	.208	.127

SOURCE: "Standard pattern" energy-consumption equation from Table 1.

a/ See text for assumptions on energy prices. The $25 figure roughly corresponds to that for the U.S. in the late 1960s while the $75 estimate attempts to reflect the higher international level of oil and coal prices which existed in mid-1974. The source of price data was the wholesale price series published in the U.N. *Monthly Bulletin of Statistics*.

TABLE 3

Costs of Energy Relative to Export Earnings, Various Income Levels, Small and Large Countries, Assuming Common Development Patterns and Energy Prices

GDP per Capita, US $		"Standard Export/GNP Ratios[a]		Increase in Export Earnings Needed to Finance Higher Energy Costs Entirely from Exports[b]	
1964 prices	1970 prices	10 mil. pop.	100 mil. pop.	10 mil. pop.	100 mil. pop.
100	125	.215	.106	24%	27%
200	250	.250	.100	34%	52%
400	500	.276	.100	46%	77%
800	1000	.294	.107	51%	86%
1600	2000	.303	.122	49%	75%
3200	4000	.303	.142	44%	58%

[a] Computed from large and small country export equations found in N.G. Carter, "A Handbook of Expected Values of Structural Characteristics," Bank Staff Working Paper 154, June 1973, I.B.R.D. The equations are identical to those reported by Hollis Chenery and Moises Syrquin, Patterns of Development 1950-1970 (London: Oxford University Press, 1975), Tables 39 and 510, pp. 204, 205. In using the equations, net capital inflows or outflows from the country were assumed to be zero.

[b] See Table 2 for the energy/GDP ratios from which these percentage increases were derived. The increases were assumed to result from a rise in average energy cost from $25 to $70 per MTCE.

capita and that there is some tendency among both smaller and larger countries for relative export earning to increase. This reduces the relative problem of energy supply and financing, but only by modest amounts. For a country such as Japan with a 1970 population of 103 millions and a per capita GNP in the range of $2000 at official exchange rates, energy costs at 1970 prices might have amounted to about 5 per cent of GDP. Under the recent price increases, Japan, with three-fourths of its fuel from petroleum, might find itself allocating close to 15 percent of its GDP to energy purchases. (Normal export ratios for a country the size of Japan are about 12 percent of GDP, but the actual ratio for Japan in recent years has been about 10 percent. This probably relates to the fact that Japan has been a net exporter of capital, a factor which reduces the standard exports of merchandise in the statistical relationship.)

The conclusions of this section are that energy costs have recently risen to significant fractions of national product and that these costs can be expected to rise as a fraction of GDP during the early and middle stages of economic development. The magnitudes are such to provide strong incentives to newly developing countries to (a) exploit whatever energy sources they may themselves possess, (b) consider development patterns aimed at reducing the net costs of energy, after allowing for whatever increases in capital costs may accompany the chosen alternative, and (c) to the extent to which energy will not be available from domestic sources, encourage a faster-than-normal growth of foreign exchange earning exports to pay for the anticipated increases in energy imports.

Components of Energy Demand

A considerable amount of research has been undertaken in recent years to identify energy links with selected economic activities and functions. Such studies provide useful insights about energy use in the more advanced countries, but little or not work has yet been conducted on parallel studies in developing countries. No system for allocating energy use can be ideal for every purpose, and in the current instance a simple four-way breakdown has been chosen. The four proposed categories of energy use are:

1. Substitution for human labor in fields and factories.
2. Production of materials from natural resources, including the production of additional amounts of energy.
3. Assistance with the organization and acceleration of the production process.
4. Contributions to the "quality of human life," not included under 1.

The first category of energy use, while traditionally held to be of first importance to human welfare, is quantitatively minor and increases over time more slowly than does GDP. The energy value of the food eaten by India's masses (1990 Kcal/per person per day) amounts to a coal equivalent of about 106 KCE per person per year. It is doubtful, however, that more than 2.5 percent of this results in useful work. (This assumes that about one-half the population is employed at physical labor and that the energy conversion efficiency of the human body is about five percent.) If machines could be used to replace human labor at a system-wide energy use efficiency of 10 percent, replacement inanimate energy might amount to 36 KCE/person/year or approximately 14 percent of India's total commercial energy use of 192 KCE/person in 1970. This human-labor-substitution energy could be expected to increase over time only slightly faster than population growth and would drop to less than 10 percent of total commercial energy use for a country with an income level of Brazil (about US $400 per person per year at official exchange rates).

Table 4 presents data on material production and attendant energy costs for six representative countries: India, Pakistan, Brazil, Italy, Japan, and the U.S.A. Energy use factors are based upon recent U.S. experience. (See Annex Table I for the derivation of the energy-use estimates.) Commodities are limited to intermediate, "bulk" materials in order to minimize energy double-counting. Energy for producing the ores, etc., needed for bulk material production is explicitly included in the direct-plus-indirect energy use coefficient (column 7 of Annex Table I) used to compute total energy requirements attributable to each bulk commodity.

Table 4 was intended to demonstrate that the more exotic, energy-intensive materials become increasingly important in wealthier countries (and probably also, by inference, during the course of economic development). The net effect would be to show that consumption of energy for material production increased at a rate faster than manufacturing as a whole, perhaps even exceeding the growth rate of total commercial energy use. The assumption was that many U.S. industrial processes have become more efficient over time with respect to energy use. Weighing material

TABLE 4

Energy Required under U.S. Technology[a/] for the Production of Selected Commodities, Various Countries, 1969-71

	India	Pakistan[b/]	Brazil	Italy	Japan	U.S.A.
I. GROSS TOTAL ENERGY, Btu x 10^{12}						
Refined petroleum products	96	24	139	616	922	3032
Pulp and paper	27	4	35	61	366	1597
Chemicals[c/]	52	4	65	338	1225	2490
Cement, hydraulic	125	23	78	242	493	625
Steel, basic	322	5	320	914	4657	6306
Non ferrous, light[d/]	38	--	15	38	182	904
Non ferrous, other[e/]	3	--	4	15	148	326
TOTAL, Btu x 10^{12}	663	60	656	2224	7993	15280
TOTAL, coal equivalent, MTCE x 10^6	24.29	2.20	24.03	81.47	292.8	559.7
II. TOTALS, COAL EQUIVALENT, PER:						
a. Inhabitant (KCE)	44	17	252	1517	2832	2733
b. Unit total commercial energy (MT/MT)	.230	.172	.431	.537	.852(!)	.240
c. Manufacturing						
1. Official exchange rates:	3.51	.75	3.20	2.78	4.10	2.19
2. Purchasing power deflated:	1.02	.33	1.55	2.14	2.79	2.19

a/ Direct and indirect requirements in the primary and all supplying industries. Assumes U.S. technology, industrial structure, degree of fabrication (product mix), and relative dependence on imported intermediate products. Energy coefficients are derived in Annex Table I, and detailed production figures are shown in Annex Table II.

b/ Including what is now Bangladesh.

c/ Including fertilizers and insecticides but excluding basic industrial chemicals as usually defined (ISIC 3511). Also excludes aluminum oxide used in production of aluminum.

d/ Aluminum and magnesium.

e/ Copper, lead, and zinc.

production by U.S. energy-use coefficients, therefore, would if anything impact a downward bias to the results and therefore strengthen the conclusion.

Several complicating factors in Table 4, however, should be noted. The Japanese energy total appears improbably large, amounting to 85 percent of total energy consumption in the country. This suggests that energy-use efficiency may be considerably less in Japan than in the U.S., rather than theother way around as had been originally assumed.[5/] Second, a number of energy-intensive materials-producing sectors, principally basic industrial chemicals, contain large numbers of commodities and thus can be but sketchily covered in international compilations of production statistics. The effect is probable to bias downward U.S. materials-associated energy use in comparison with India and Brazil. On the other hand, the most common form of energy-intensive building material in the poorer countries, lime and sun- and locally kiln-baked bricks, are also imperfectly covered, thus providing a bias in the opposite direction. (Much of the energy to produce lime and bricks in India, Pakistan, and Brazil, however, probably comes from "non-commercial" energy, principally fuel wood.)

Finally, inter-country differences in manufacturing value added are probably unduly exaggerated when measured at official exchange rates. Table 4 therefore shows an alternative measure based upon the purchasing-power equivalent GDP deflators discussed earlier.

The conclusions derived from Table 4 appear to be as follows. If the high energy use estimate for Japan is regarded as an anomoly, energy use for the production of the major energy-intensive materials declines in relationship to manufacturing value added, measured at official exchange rates, as per capita GDP increases. Making a rough correction for purchasing power differences among the countries, however, reverses this conclusion. Furthermore, if an allowance could be made for energy consumed in producing industrial organic and inorganic chemicals (other than fertilizer and plastics), it is probable that an increasing portion of commercial energy would be used for material production at higher levels of per capita GDP. There is a tendency for the consumption of almost all materials to begin falling in proportion to GDP at some point,[6] however, and it may be that Japan is less of an anomoly than first appears. It is possible that commerical energy for materials production reached a peak (as a fraction of total commercial energy use) at some intermediate stage of economic development and declines thereafter.

For countries in the early stages of development, however, the conclusion appears quite clear. The production of the bulk materials which form the building-blocks of modern industrial societies (steel, aluminum, copper, cement, plastics, paper, and refined petroleum products) require large amounts of energy. One-fourth of India's commercial energy use may be allocated already to the production of the commodities listed in Table 4, assuming that Indian production processes are as efficient as the U.S. , and this fraction of the total will probably increase as industrialization proceeds. If Indian production processes use less energy on a system-wide basis than the U.S., as might occur if ancilliary activities that are mechanized in the U.S. are more labor-intensive in India,[7] then the future increase in energy use could be even more rapid.

This conclusion may be further highlighted by examining the price implications of Table 4 and Annex Table I. The direct and indirect energy "content" of the commodities covered by Table 4 ranges from 9.4 KCE per dollar for the commodities produced by India down to 7.8 KCE for those produced by the U.S. (that is, the mix of the commodities shown becomes slightly less energy intensive per dollar of output, at U.S. technology and prices, as one moves from the low income to the higher income countries). At an average energy price to industry of about \$17/MTCE,[8/] this would suggest that direct and indirect energy cost as a portion of final selling price might have been about 16% for India in 1970 and 14% for the U.S. If the recent three-fold increase in energy prices is fully reflected in prices paid by industrial consumers, this would indicate the need for a 30% average increase in final selling prices for these more energy-intensive goods in India and perhaps 24% in the United States. At the higher final selling prices, direct and indirect energy costs would be equal to about 36% of the total in India and to 31% in the United States.

Thus fuel and fuel-using efficiency becomes of major importance, under the recent energy price changes, in the cost structure of energy-intensive industries. Although the traditional development sequence appears to have been accompanied by a relatively rapid increase in the production of this group of commodities during the early stages of growth, developing countries in the future may wish to consider alternatives to this increasingly costly traditional pattern.

Energy used elsewhere in the production process (and to some extent already included in the _indirect_ energy requirements for materials production) is largely for the physical workings, shaping, transporting, assembling, and marketing of various manufactured goods. Important aspects are the creation of new productive capacity and the organization of the productive process (through communication and transportation) so that existing capacity is utilized to its fullest extent. This category of energy use often involves the speeding up of productive processes. Thus tractors may be justifiably used in a labor-surplus economy because they shorten land preparation time and permit more crops to be grown in a particular time period. Other energy use has more to do with the efficient organization of the productive process.

Energy accounts do not usually distinguish between bulk material production and materials working, assembly, and marketing (and indeed the line of demarcation between the two sets of activities is not always precise). It might be supposed, however, that this latter use of energy increases at a rate equal to manufacturing production as a whole. It is possible also that as specialization increases and transportation links (including those in the marketing chains) become more important, energy devoted to finished goods production and distribution grows more rapidly than manufacturing output.

Further work is obviously required to clarify the functions performed by industrial energy as the course of development proceeds. International comparisons should be made, as a first step, based upon detailed input-output tables of several economies. To minimize problems resulting from variations in energy prices paid by different consuming sectors, energy flows in these tables should be measured in physical units and broken down as much as possible into the various forms of fuel consumed. Since transportation is one of the majore energy-consuming sectors, special attention should be paid to transportation margins on delivered goods and to the energy costs of this transportation.

The fourth energy use category suggested earlier is that of consumption to improve the quality of human life, other than energy consumption which directly substitutes for human labor. Energy in this fourth category is largely consumed directly by final consumers (households) and includes space heating (residential and possibly some commercial), water heating, air conditioning, refrigeration, electric lighting, energy for household appliances, and fuel for most passenger transportation. To some extent the demand for this type of energy depends upon climate, but personal disposable income probably plays a dominant role. The effect of prices is less certain, but there has undoubtedly been a tendency for lower energy prices (relative to other goods and services) to contribute over a period of years to increased consumer demand for energy and for the complementary energy-using equipment. Even for such a relatively fixed commodity as space heat, studies in the U.S. have shown a significant income effect (although the income elasticity is less than 1.0) and a smaller and possibly less significant price effect.[9/]

Little is known about this category of energy use in developing countries except that in the poorest countries it is probably modest and derived largely from non-commercial fuels. Arjun Makhijani and Alan Poole, in an innovative study of energy use in rural areas, calculate rural cooking needs at about .3 tons of wood[10] (200 KCE) per inhabitant and space heating consumption in China at a further fuel wood consumption of perhaps 1.3 tons (800 KCE) per capita.[11] These estimates dramatize the importance of so-called non-commercial fuels which in both India and Brazil may be equal in heat content to total commercial energy consumption.

The conclusion of this section is that in terms of the four functional categories defined, we can expect the following changes to take place during the traditional (historical, or pre-energy=price-increase) development sequence:

1. Energy which substitutes for human labor, while very important for improvements in the quality of life, is quantitatively small and becomes relatively insignificant as development proceeds.
2. Energy for the comfort and convenience of households, including personal transportation, is probably quite large initially as a percentage of total energy use. As development proceeds it is conceivable that this energy shows little or no increase per capita during early growth stages as more efficient energy-using equipment becomes affordable. The usual pattern, however, has been for a gradual shift to commercial fuels, thus often creating a disproportionately large demand on foreign exchange.
3. Use of fuel and power, directly and indirectly, to produce a relatively small group of energy-intensive bulk commodities, may be expected to increase about as fast as the manufacturing sector as a whole, as conventionally measured. Energy for bulk material production is thus likely to constitute a rapidly

increasing share of total commercial energy.

4. The residual category of energy use covers the processing of bulk materials into finished products and their marketing and distribution to intermediate and final consumers. The energy-using processes largely involve mechanical effort that is less energy-intensive (in terms of Btu/dollar of output) than for bulk material manufacture. It is possible that this category of energy use grows less rapidly than manufacturing as a whole, in the historical development sequence, although probably at least as fast as gross domestic product.

Implications for Research

It should be obvious from the preceeding discussion that we know very little about how and where energy is used in the world. Most of the speculations in this paper are derived from very broad, economy-wide comparisons among countries, from more detailed knowledge of the world's richest society, the United States, and from the sketchiest kinds of reports on one or two countries at the other end of the economic spectrum. For a much broader range of developing countries we need detailed, basic knowledge of what fuels are used for what purposes, in what locations and at what efficiencies of use. This knowledge should be organized in terms of functional systems of capital and energy. From the energy using viewpoint, the four functional categories proposed in this paper appear to have merit: conditions of human labor, comfort/convenience of households, production of bulk materials, and materials using

TABLE 5

Household Use of Energy, U.S., 1960, 1968, and 1970

		1960	1968	1970
I.	Residential[a]	Consumption, as % of national total		
	Space heat	11.3	10.9	10.8
	Water heating	2.7	2.9	2.9
	Cooking	1.3	1.1	1.0
	Clothes drying	0.2	0.3	0.3
	Refrigeration	1.7	1.6	1.6
	Air conditioning	0.2	0.3	0.3
	Other (small appliances, lighting, etc.)	1.1	2.1	2.4
	SUBTOTAL	18.5	19.2	19.3
II.	Passenger automobile use[b]	11.6	11.6	12.0
III.	TOTAL %	30.1	30.8	31.3
IV.	Energy use, heat-value equivalents		Physical Units	
	A.1. Total, US Btu x 10^{12} c/	44,569	61,700	67,143
	2. Coal equivalent, UN MTCE x 10^6 d/	1,506	2,086	2,269
	B.1. Residential and passenger automobile, MTCE x 10^6 e/	453.4	642.5	710.4
	2. Per capita, KEC/person f/	2510	3200	3470

a/ Stanford Research Institute, Patterns of Energy Consumption in the United States, p. 16, Table 3. 1970 percentages extrapolated from 1960-1968 trends. Compare with Eric Hirst and John C. Moyers, "Efficiency of Energy Use in the United States," in Philip H. Abelson, ed., Energy: Use, Conservation, and Supply (Washington: American Association for the Advancement of Science, 1974), p. 14, Table 1.

b/ Derived from passenger car mileage and gas consumption data (assuming 5.3 Btu x 10^6 per barrel of gasoline) in Hans H. Landsberg, "Low-Cost, Abundant Energy: Paradise Lost?" ibid., p. 4, Table 4.

Continued on following page

[c] U.S. Bureau of Mines estimates. See <u>Statistical Abstract of the U.S.</u>, 1973, pp. 508-509.

[d] U.N., World Energy Supplies, 1969-1972, p. 13, Table 2, for 1970 estimate. The 1960 and 1968 estimates are based on line IV.A.1 and an implicit heat equivalent for coal of 29.58 Btu x 10^6/MT. (This latter appears high but is consistent with the two different data sources for 1970.)

[e] Line IV.A.2 x Line III ÷ 100

[f] Line IV.B.1 ÷ population

and distribution activities, including the creation of additional productive capacity.

From the viewpoint of any developing country <u>not</u> sitting on a pool of oil, the most fundamental question is how to expand domestic energy sources. Many technological possibilities become economically interesting at current fuel prices, and there appears to be the expected proliferation of development research on solar, wind, geothermal, and vegetal energy sources. These tend to be lower grade but more widely dispersed than traditional fossil fuels and so of particular interest to energy-poor developing countries. This research should, of course, be encouraged and should be linked where possible to broader questions of future life styles, patterns of industrial location, and the international division of labor.

A second set of considerations for a developing country concern likely changes in future energy demand during the course of development. This paper sets forth some hypotheses about how demand may have evolved under the low-cost energy pattern of the past. Little is said, however, about alternatives which developing countries might consider under today's much higher prices. Little <u>can</u> be said, in fact, until we understand more fully the roles which energy-capital systems play in the development process. Many of the questions involved are those which have perplexed developmental economists for many years. By focusing on the technological as well as the economic basis for development, however, we can perhaps gain a clearer appreciation of both energy use and economic growth itself.

Three broad problem areas would seem to be of particular interest:

1. What kind of energy supply and delivery systems are feasible and economic for widely dispersed populations such as usually characterize rural areas? Wood and vegetable wastes have been the traditional fuels in these areas, gradually giving way to kerosene, gasoline and rural electrification as incomes rise. There is renewed interest today, however, in solar, wind, liquified biogas, or some combination of these and conventional energy systems. (These are the systems considered by Makhijani and Poole in the book cited several times earlier in this paper.) A next step would be field testing of alternative systems, and it is reported that Makhijani has made a proposal for at least one such test in India. A basic problem, however, may be that economic development as it has so far occurred in the Western world has been geared to and dependent upon moving people out of agricultural and rural areas and into population concentrations which could support "modern" industrial activity. Most well-meaning efforts to reverse this trend and decentralize industry have been failures. Today, considerations of social justice as well as the logic of huge rural populations in many of the poorer countries have created new interest in "rural development" - in putting capital investment into rural areas rather than in exploiting farmers for the benefit of the politically more potent urban masses. The question is whether today's higher energy costs and the various technological responses under consideration will make it easier or more difficult to

bring about the contemplated revolution in the rural/urban division of economic gains.

2. What are the trade-offs among energy, capital equipment, production time, and product characteristics in the production of bulk materials? Is it a technological necessity that future production of many of these materials will be as energy-intensive as appears to be the case today in the United States? What alternative productive processes and systems might be considered which would be more appropriate under today's relative price structure? What possibilities exist for greater international specialization (and efficiency) in the production of bulk materials? And in this connection, how important appear to be the technical, industrial, and other linkages which have led countries to believe that domestic production of most of these commodities is an essential step in their own industrialization?
3. What are the amounts consumed and the functions performed by energy in cities? This is where most of the material-using and related productive activities of economies take place, and there is strong evidence that the urbanization process itself has been causally linked with industrial growth. As with bulk-materials production, we would like to know the ranges for substitution among capital, energy, and time, and whether patterns that evolved in the low-cost energy past will continue to make economic sense in a higher-cost energy future. Since a city is a moderately self-contained unit with defined boundaries and

relatively limited numbers of exit and entry points, it should be possible to study urban energy systems just as biologists have studied other ecological systems. A comparative study of 4-5 cities in countries at different stages of development should provide a great deal of information both on the materials-using and production-organizing aspects of growth as well as on the comfort/convenience aspects, which have tended to become most fully developed in urban areas.[12]

Footnotes

1. This paper was written for the MIT Energy Laboratory as part of an effort to explore the links between energy use and economic development.

2. The standard energy-using relationship shown in Table 1 was based upon a sample of 31 more-urbanized countries, including Mainland China and four other eastern bloc countries, whose populations in 1970 ranged from 10 millions to 760 millions. Six countries had populations above 100 millions, and seven had populations of 20 millions or less. See the notes to Table 1 for details of the statistical relationship itself.

3. A System of International Comparisons of Gross Product and Purchasing Power (Baltimore: Johns Hopkins, 1975).

4. The statistical relationship found was:

$$DEFL = 46.34 + .130\ GDP/POP - .0338\ POP \qquad \overline{R^2} = .925$$
$$(13.41) \quad (9.88) \qquad (2.80) \qquad \text{See as \% of mean: } 8.9\%$$

where DEFL = purchasing power GDP deflator estimated by Kravis et al (page 9, Table 1.5)
GDP/POP = Gross Domestic Output per capita, 1970, U.S. dollars, from UN Monthly Bulletin of Statistics, January 1975.
POP = population in millions, ibid., November 1971. and t-ratios are shown in parentheses.

5. This probably relates to the more modern steel industry in Japan, higher rates of capacity utilization, and the greater use of top-blown oxygen converters for making steel (about three-fourths of Japanese production in contrast to one-half in the United States).

6. This conclusion is based on a series of cross-country materials-consumption regressions along the lines of that used for the standard energy consumption pattern of Table 1. The eventual decline-in-proportion-to-GDP, however, appears largely related to the conventional measurement of GDP at international exchange rates. When the crude adjustment for purchasing power difference - discussed earlier - is made, steel consumption shows an eventual decline (i.e., a statistically significant negative coefficient for the squared-GDP per capita variable), but alumimum and copper do not.

7. The basic raw materials-transformation processes are probably similar in the two countries with the U.S. possibly having the edge in energy-use efficiency because of scale effects and more sophisticated technology. System-wide energy use, however, involves many supporting activities such as materials handling both in the industry of primary production and in industries which supply the primary producer. Direct-plus-indirect energy-use coefficients for the U.S., for example, may be several times the direct coefficients in the primary pro-

ducing sector. Compare, in this connection, the estimates shown in Annex Table I with those calculated by the Stanford Research Institute (in "A Study of Process Energy Requirements for U.S. Industries") and reported in Stanford Research Institute, Patterns of Energy Consumption in the United States, Office of Science and Technology, Executive Office of the President (Washington: U.S. Government Printing Office, January 1972), p. 152, Table 69. (The Herendeen and Bullard study used as a basis for Annex Table A-1 did not report direct energy use coefficients.)

8. By analogy with the structure of Btu prices in the United States in 1967. In that year industrial consumers (excluding electric utilities but including the full thermal-generating heat value of electricity purchased by other industrial consumers) probably paid about half as much for energy, on a Btu basis, as did residential consumers. The industrial unit price was perhaps about 70 percent of that for the economy as a whole, excluding electric utilities and other transformed types of fuel and power. The $17 per MTCE estimate for 1970 is based on an economy-wide average price of $25 per MTCE. As noted in the earlier part of this paper, it has been assumed for the purposes of comparison that the higher oil prices have not yet been fully reflected in higher prices for coal and that an economy-wide average price in 1974, assuming that international prices generally prevailed throughout the economy, might have been in the neighborhood of $70 per MTCE.

9. Alan M. Strout, "Weather and the Demand for Space Heat," Review of Economics and Statistics, vol. XLIII (May 1961). The income elasticity found in this study was .78 (± .14) and that for price, -.27 (± .17). The relationship covered the years 1935-40 and 1946-51.

10. Makhijani and Poole, Energy and Agriculture in the Third World, A Report to the Energy Policy Project of the Ford Foundation (Cambridge, Mass: Ballinger, 1975), p. 69.

11. Ibid., p. 71. Elsewhere (p. 39) the authors report that wood consumption in Gambia and Tanzania averages, respectively, about 1.0 and 1.5 tons per capita (630 to 950 KCE) and that per capita energy use for cooking in rural areas is 6 to 7 million Btu per year (210 to 260 KCE). Energy from wood "supplies about 10 times more energy than all commercial fuels" in these two countries.

12. The beginning step in such a study should be a careful reading of Richard L. Meier, The Design of Resource-Conserving Cities (Cambridge, Mass: M.I.T. Press, 1975.) For examples of studies of other energy-using systems, see the several articles published in the September 1971 issue of the Scientific American and reprinted under the title Energy and Power (San Francisco: W.H. Freeman and Company, 1971).

ANNEX TABLE I

Direct and Indirect Primary Energy Use per Physical Output of Products, Selected Energy-Intensive Materials, United States, 1967

Industry Input-Output No.	Industry S.I.C. No(s)	Name	Direct and Indirect Primary Energy Use per Unit of Final Output	Manufacturer's Price, $ per ST	Direct and Indirect Btu X 10^6 per ST	Direct and Indirect Btu X 10^6 per MT
(1)	(2)	(3)	(4)	(5)	(6)	(7)
3101	2911,299	Petroleum refinery and products	1.2082 Btu/Btu	---	5.2 a/	5.7 a/
2401	2611	Pulp mills	200,551 Btu/$	$120	24	27
2402	2621	Paper mills	201,228 Btu/$	229	46	51
2403	2631	Paper board mills	219,213 Btu/$	121	27	29
2406	2644,2661	Building paper and board	190,326 Btu/$	115 b/	22	24
--	--	Paper and paper board	--- Btu/$	---	34	38
2702	2871,2872	Fertilizer	173,931 Btu/$	112 d/	19	21
2703	2879	Agricultural Chemicals	167,757 Btu/$	536	90	99
2801	2821	Plastic materials and resins	216,753 Btu/$	616	134	147
2802	2822	Synthetic rubber	293,302 Btu/$	460	135	149
2803	2823	Cellulosic man-made fibers	208,117 Btu/$	925	201	221
2804	2824	Organic fibers, non-cellulosic	141,607 Btu/$	1654	234	258
3601	3241	Cement, hydraulic	481,161 Btu/$	16.6	8.0	8.8
3602	3251	Bricks and structural clay tile	340,560 Btu/$	18.6 e/	6.3	7.0
3701	331	Steel products f/	267,425 Btu/$	174	47	51
3801	3331	Primary copper	139,706 Btu/$	(794)774 g/	(111)186 j/	(122)205
3802	3332	Primary lead	110,162 Btu/$	251	28	30
3803	3333	Primary zinc	274,427 Btu/$	(273)242 h/	(75)81 j/	(83)89
3804	3334,28195	Primary aluminum	387,646 Btu/$	236 i/	222 j/	245

See notes on following page.

Annex Table I, continued

Col (1), (2): Survey of Current Business, vol. 54 (February 1974), page 35.

Col (4): Robert A. Herendeen and Clark W. Bullard III, "Energy Costs of Goods and Services, 1963 and 1967," CAC Document No. 140, Center for Advanced Computation, University of Illinois at Urbana - Champaign, November 1974, Table 4b.

Col (5): Except where U.N. sources are indicated, from U.S. Bureau of the Census, Census of Manufacturers, 1967, Vol. II, Industry Statistics (Table 6A), (U.S. Government Printing Office, 1971).

Col (6): Equals columns (4) x (5) except where noted.

Col (7): Equals column (5) x 1.1023 Short ton/metric ton (before rounding column 5)

a/ Excludes Btu equivalent of the refined petroleum output itself.

b/ Excludes wallpaper (SIC 2644) for which weight is not available.

c/ Weighted average of SIC 2621, 2631, 2661.

d/ In terms of nutrient content (NPK) as reported by the United Nations, Growth of World Industry, 1972 ed.,Vol. II, Commodity Production Data, 1963-1972 (New York: United Nations, 1974).

e/ Assumes that the density of bricks is two times that of water. The usual measure of bricks is either in terms of million standard bricks or cubic meters (One million standard bricks = 1019 M^3).

f/ This sector includes pig iron production, coke-oven operated by steel companies, steel works, rolling mills, electrometellurgical products, wire products, cold rolled steel, and pipes and tubes. Energy is used in all phases but the relatively unduplicated output of the industry has been taken to equal:

	1967 (UN Data)
Pig iron for foundries	3403 MT x 10^3
Other ferro alloys	1656
Crude steel for casting and crude steel ingots	115406
TOTAL	120465 MT x 10^3

Energy used directly and indirectly throughout the entire iron and steel making-and-forming sector is thus expressed in terms of (roughly) "crude steel equivalent."

continued on following page

Annex Table I continued

g/ The output of this sector includes about 42% by weight of copper smelter products (non-refined) and matte. The price of refined copper alone was $794/ST.

h/ Including 18% by weight of zinc residues and smelter products. The price of refined zinc was $273/ST.

i/ Including 59% by weight of aluminum oxide (SIC No. 28195). The price of the refined aluminum was $475/ST.

j/ Estimated by allocating the entire amount of direct and indirect energy attributes to the sector to the amount of refined metal. Where a part of this output was refined from imported intermediate products, the total (domestic and foreign) energy use will be understated. On the other hand, the coefficient will overstate domestic energy requirements, should imported intermediate products account for a larger share than in 1967. According to the U.S. Bureau of Mines, these 1967 shares were

	% refined metal derived from foreign ores, mattes, etc.	Source: Statistical Abstract of the U.S., 1969, page:
Copper	25%	680, Table 1051
Zinc	53%	681, Table 1053
Lead	32%	682, Table 1054
Aluminum	28%	Assumes two tons alumina (aluminum oxide) per one ton refined aluminum. Of the calculated requirements, for 8258 ST x 10^3 alumina, 5917.4 ST x 10^3 was reported produced by the U.S. chemical industry.

ANNEX TABLE II

Production of Selected Energy-Intensive Bulk Commodities: Six Countries, 1969-1971 Average (Where Available) (MT x 10^3)

Commodity (1)	ISIC No. (2)	Direct & Ind. Primary Energy per MT (3)	India (4)	Pakistan (5)	Brazil (6)	Italy (7)	Japan (8)	USA (9)	Source
Petroleum refinery products	3530	5.7	16922	4130	24346	108124	161669	531950	A
Wood pulp	3411 part.	27	656	92	867	889	8508	38645	B
Paper and paper board	3411 part.	11 e/	819	129	1056	3388	12397	50356	B
Fertilizer, MPk equivalent	3512	21	940	123	289	1704	3223	15126	C
Insecticides, etc.	251216	99	---	---	127	---	25	513	C (note a)
Plastic materials and resins	3513	147	10	---	66	1157	4658	8123	C (note b)
Synthetic rubber	351301	149	28	---	98	154	668	2265	C
Cellulosic, man-made fiber	351307,40	221	101	5	50	191	508	656	C
Organic fibers, non-cellulosic	351304,37	258	16	2	44	260	1000	1725	C
Cement, hydraulic	369204	8.8	14149	2642	8877	32097	56012	70974	F
Building bricks	369101	7.0	2156	(n.a.)	8405	19636	(n.a.)	15034	C (note c)
Steel products	(noted)	51	6320	100	6274	17921	91304	123642	C
Primary copper	372001,4(?)	122(?)	10	---	18	13	683	1941	D
Primary lead	372037(?)	30	2	---	21	78	204	729	D
Primary zinc	372043	83	23	---	10	137	703	814	E
Primary aluminum	372022(?)	245	157	---	60	142	727	3536	D
Primary magnesium	(?)	370 f/	---	---	---	8	10	101	D

See following page

ANNEX TABLE II continued

n.a. = not available; a dash may also indicate not available rather than zero.

SOURCES:

A. U.N., World Energy Supplies, 1969-1972, Statistical Papers Series J, No. 17, (New York, 1974), Table 10.

B. Pulp and Paper, World Review Issues, 1971, 1972, 1973 (Miller Freeman Publications, San Francisco).

C. U.N., Growth of World Industry, 1972 edition, Vol. II (New York, 1974).

D. Metal Statistics, 1963-1973, 61st edition (Metallgesellschaft Aktiengesellschaft, Frankfurt am Main, 1974).

E. Yearbook of the American Bureau of Metal Statistics, 53rd Annual Issues, 1973.

F. Statistical Summary of the Mineral Industry, 1967-1971 (London: Her Majesty's Stationary Office, 1973)

NOTES:

a/ Shipments covered appear to account for only about 33% (1967 data) of SIC sector 2879 (Agricultural Chemicals, nec) in the United States.

b/ Includes ISIC numbers 3513-10, -13, -16, -19, -22, -25, -28, and -31. Total may have represented 72% of the comparable U.S. manufacturing sector in 1967.

c/ Converted to metric tons using factors of:
2038 metric tons/million standard bricks
or 2.0 metric ton/cubic meter of bricks

d/ Consists of pig iron, foundry ISIC No. 371007
other ferro alloys 371013
crude steel for castings 371016
crude steel ingots 371019

The assumption is that all other pig iron is used for the production of steel and therefore should not be included in an unduplicated total.

e/ To give an energy use total which is approximately unduplicated, the paper and paper board coefficient has been reduced by the amount of energy used to produce pulp.

ANNEX TABLE II continued

f/ Not based upon input-output calculation but upon direct energy use coefficients from Stanford Research Institute, Patterns of Energy Consumption in the United States, Office of Science and Technology, Executive Office of the President (Washington: U.S. Government Printing Office, 1972), p. 152, Table 69. A comparison with the input-output derived coefficients from Annex Table II, suggests that the SRI coefficient for magnesium should be multiplied by a factor of four to approximate total direct and indirect energy requirements:

	SRI direct coefficients	Input-output, direct-plus-indirect cost	Ratio
Aluminum	60.8	245	4.02
Magnesium	92.6	n.a.	n.a.